T0328310

Sustainable Mega City Communities

Sustainable Mega City Communities

Edited by

Woodrow W. Clark II, MA³, PhD

Founder/Managing Director
Clark Strategic Partners
Beverly Hills, CA, United States

Butterworth-Heinemann
An imprint of Elsevier

Butterworth-Heinemann is an imprint of Elsevier
The Boulevard, Langford Lane, Kidlington, Oxford OX5 1GB, United Kingdom
50 Hampshire Street, 5th Floor, Cambridge, MA 02139, United States

Notices
Knowledge and best practice in this field are constantly changing. As new research and
experience broaden our understanding, changes in research methods, professional
practices, or medical treatment may become necessary.

Practitioners and researchers must always rely on their own experience and knowledge in
evaluating and using any information, methods, compounds, or experiments described
herein. In using such information or methods they should be mindful of their own safety
and the safety of others, including parties for whom they have a professional
responsibility.

To the fullest extent of the law, neither the Publisher nor the authors, contributors, or
editors, assume any liability for any injury and/or damage to persons or property as a
matter of products liability, negligence or otherwise, or from any use or operation of any
methods, products, instructions, or ideas contained in the material herein.

Library of Congress Cataloging-in-Publication Data
A catalog record for this book is available from the Library of Congress

British Library Cataloguing-in-Publication Data
A catalogue record for this book is available from the British Library

ISBN: 978-0-12-818793-7

For information on all Butterworth-Heinemann publications visit our website at
https://www.elsevier.com/books-and-journals

Publisher: Matthew Deans
Editorial Project Manager: Gabriela D. Capille
Production Project Manager: Sreejith Viswanathan
Cover Designer: Alan Studholme

Typeset by TNQ Technologies

Working together
to grow libraries in
developing countries

www.elsevier.com • www.bookaid.org

Contents

SECTION 4 Globalism and regionalism: overview

SECTION 5 Conclusion

Contributors

Danilo Bonato, MBA
General Manager, ReMedia, Via Messina 38, Milano, Italy

Elisa Castoro
Professor, Graduated in Foreign Languages and Literatures, English Translator and Photographer, Matera, Italy

Serra Çelik, PhD
Professor, Informatics Department, Istanbul University, Istanbul, Turkey

Woodrow W. Clark II, MA³, PhD
Founder/Managing Director, Clark Strategic Partners, Beverly Hills, CA, United States

Akima Cornell, PhD
Principal, Akima Consulting, LLC, Los Angeles, CA, United States

Andrew DeWit, PhD
Professor, Kikkyo University, Toshima City, Tokyo, Japan

Michael Gartman
Professor, Poli-Technical University Milan, Italy

Murat Ozzor, PhD
Professor, Informatics Department, Istanbul University, Istanbul, Turkey

Sevinç Gülseçen, PhD
Chair and Director, Informatics Department, Istanbul University, Computer Science and Application Center, Istanbul, Tukey

Şiir Kılkış, PhD
Senior Researcher, The Scientific and Technological Research Council of Turkey, Atatürk Bulvarı, Kavaklıdere, Ankara, Turkey

Fatma Önay Koçoğlu, PhD
Professor, Informatics Department, Istanbul University, Istanbul, Turkey

Lucia Elsa Maffei
Private Law Practice, Lawyer qualified to the higher Courts, Matera, Italy

JudiGail Schweitzer-Martin, MRED, AMDP, CGBP, CALGreen CAC, ENV SP, SBE/DBE
Adjunct Professor, University of California, Irvine President — Chief Sustainability Advisor, Schweitzer + Associates, Inc., Lake Forest, CA, United States

Wang Weiyi, PhD
Associate Professor, Jiaxing University, Zhejiang Province, People's Republic of China

Yueqi Zhou
EMBA Graduate, Cheung Kong Graduate School of Business, Beijing, China

Introduction

Woodrow W. Clark, II MA³ PhD

Climate change is real

The facts are overwhelming now more than ever, as the UN IPCC (United Nations, Inter-governmental Panel on Climate Change, 2019) report documents over the last 5 years. Additionally, the population around the world has grown while older people become healthier and live longer. Hence retirees move away from areas and states where climate has changed dramatically and thus are too dangerous for them and their loved ones to live due to the extreme weather in areas like Florida and other southern states as well as many other US states.

California is the global laboratory for the agile energy system and the first "sustainable nation-state" that is building its energy system on a new set of economic assumptions after being blacked out by deregulation built on the old assumptions. To fully develop the agile energy system, Californians must continue to put civic concerns over private profits, and build an energy infrastructure based on what is good for the public. As a "bellwether" nation-state, California has the opportunity to lead the world in the new energy system.

The energy crisis in California was a challenge for all its citizens. The "design flaws" or "restructuring," as some economists now label it, were not the only problems. As new energy systems are envisioned and constructed to respond to the crisis, policy makers must reformulate the basic premises that led to the crisis in the first instance. This means reevaluating the political economic foundations that led to deregulation. These assumptions must be recast to provide a new direction that will provide electricity that is cheap, reliable, and environmentally friendly without relying on price competition as the main economic tool.

The extent to which the old basic premises need to be replaced is clear not only from the failure in California but also from the more widespread problems with deregulation or privatization in other states and nations. Growing evidence appears to indicate serious problems with energy sectors worldwide where unrestrained competition has been tried (Kapner, 2002), and this is not just a minor adjustment but a huge mistake. The energy crises during the summer of 2003 in Northern United States and Southern Canada along with those in Europe point dramatically to something being wrong with the deregulation and privatization economic models.

The current energy crisis created a challenge which provides the opportunity to look at energy economics in a new and different manner. While neoclassical or conventional economic theory looks at energy from the perspective of the market, energy economics needs to be examined from the perspective of the society in general. The object of an energy system or sector should not be to maximize corporate profits, but to assure that civic interests are protected for all citizens and best developed for future generations.

In the current predominant deregulation model, the pursuit of profit is assumed to lead to public good. As evidence mounts, the pursuit of the public good when it leads to profit only is a disaster for not only the company, but also the general public. This "public good" argument is parallel to Hawkin and Lovin's (1999) concept of "natural capitalism" rather than neoclassical economic theory. It is consistent with the findings that socially and environmentally responsible firms are profitable, sometimes more profitable than average (Angelides, 2003a).

However, this alternative economic framework is only beginning to be articulated. While it is not possible to present a complete new theory of civic capitalism here (Clark and Lund, 2001), this chapter will outline the basic economic elements of a new approach to electricity structure that supports the development of an "agile energy system." "Civic markets" define the role of government and regulatory oversight which is embedded in public-private partnerships. This is not a socialist or communist model (Clark and Li, 2003).

The public good is not just maximized by central planning and control or by the elimination of private ownership.

Cooperation between the public-private sectors in the form of partnerships, collaborations, rule making, setting codes and standards, and implementing programs is the new civic market model. This approach to economics and politics is an alternative to the theory that competitive market forces would increase the public good of any nation—state. By letting private monopolies control the supply (or demand) of any infrastructure sector like energy, government opens the door for mistakes like what happened in the California energy crisis between 2000 and 2002.

The worldwide energy crisis has reinforced the basic tenant that all governments must adhere to higher standards for the public good. Leaving energy, water, environment or waste, among other infrastructure sectors, to the "market" or "competitive forces" of supply and demand was wrong in the first instance. The predictable results were private monopolies gained legal control of energy supply and generation. These market forces only replaced the publicly regulated monopolies that had supplied California with power for a century of economic growth.

The task in this chapter is to argue for a new set of tools by which the economics of the power sector can be reformulated to create new solutions and opportunities for the future of all citizens. The old neoclassical competitive model which gave deregulation to California, most of America, and now around the world, needs to be replaced with a new energy/environmental economic model that builds on networks, flexibility, and innovation. Such a new economic model is well rooted in civic markets.

A new framework for understanding: the case of energy and economics from civic society

The concept of "civic markets" is put forth now in order to highlight the differences that need to be addressed in managing a complex industry such as electricity. However, civic markets also apply to other infrastructure sectors like water, waste, transportation, and education where reliance on the market forces can be either technically relied upon or financially trusted to be honest. In addition, civic markets are likely to be in the new economy and concentrated in industries that are expanding rather than contracting.

This is most clearly seen not only in monopolist industries such as energy but also in industries involving other public infrastructure such as airlines and airports, information and telecommunications, industries with high environmental impacts such as the natural resource industries, and service industries such as health and welfare. Even industries dependent on a steady stream of innovation from university and government research labs such as pharmaceuticals, life sciences, and biotechnology are moving rapidly toward civic markets or partnerships between public and private sectors. The framework for the new economics is rapidly evolving and is reflected in a growing body of thought in politics as well as business and economics (Clark and Lund, 2001).

According to conventional neoclassical economics, companies should operate with little or no government interference. Ideally companies have no regulations and taxes, etc., but contribute to societal needs on their own. Adam Smith's (rev. 1934) concept of the "invisible hand" and more recently the Bush Administration's (2001) application of it in outlining its "energy plan," are good examples of this neoclassical economic perspective: government should not be involved in energy business activities, especially regulations. In any industry, as in any country, the argument is made there is a "balance" between supply and demand which keeps prices low due to competition among the companies for customers. It is the supply-demand balance that is the basis for all energy economics and the rational for deregulation in California (Marshall, 1998) as well as similar conventional economic justifications elsewhere in the United States and worldwide.

The energy system points out the limitations to the conventional economic models and gives priority to new concepts. Many of the contrasts between the neoclassical and civic market models are matters of degree and centrality; the civic issues are "externalities" in the current models used modern economists rather than being at the core. Civic market functions must take prominence in framing competition and market economics. Following are the main differences:

- Neoclassical economic models are based on concepts of independent firms competing to gain advantage over other firms because of efficiencies, product, technology, and price, and thus meeting the public interest because they better produce what the public wants at the lowest cost. In contrast today, we better understand that firms are in networks where innovation, efficiencies, and price are the result of the interfirm sharing and cooperation rather than simply competition.
- Neoclassical models assume private sector involvement, whereas the new system is based on an increasing number of public-private partnerships and shared responsibility between the public and private sectors. Shared ownership and management control are at the root of programs that blend the public good with private initiative.
- Neoclassical models are based on premises that markets and technological systems are largely self-regulating and that government's role is limited to protecting against market power and unfair competition by enforcing laws preventing price gouging, protecting patents, enforcing contracts, and prohibiting malicious misrepresentations or corruption, etc. In contrast, we now see an expanded role for government that goes well beyond rules to creating the context for public good in expanding markets, promoting employment, and protecting the environment.

- Neoclassical models left innovation and technological change to the marketplace, whereas the new model relies on government leadership to introduce and stabilize markets for innovations that serve the public good but which may not be in the short term private interest of market leaders.
- Neoclassical models make minimal distinction between industries where it is easy for companies to enter or leave, compared with companies in grid or network industries where control of the grid constitutes a public obligation to serve and a natural monopoly. In fact, barriers to entry in a number of industries is growing because of increased interdependency and specialized materials, information, and markets that limit participation in the industry to those already involved.

The transformation away from the neoclassical and now conventional economic model that was the basic philosophical and theoretical bases, along with a bi-partisan political agenda, and hence responsible, for the deregulation framework which led to the California energy crisis and to changes in the electricity system structures in other nations, must be discussed in some detail. It is important to understand that not all existing economic philosophy and theory are dismissed. Nor are the accomplishments of neoclassical economics in solving other industrial and business problems. Nonetheless, a full discussion (see Appendix 2 for some details from Clark and Fast (2004, 2019)) must be made.

Neoclassical economics and its conventional contemporary proponents derived from a particular economic philosophy are not appropriate for the energy and many other infrastructure sectors. Furthermore, there are other economic philosophical paradigms that lead to very different economic principles and rules (Clark and Fast, 2004). In short, the explanation of economic issues surrounding the electricity industry requires new tools, frameworks models based upon a different social science philosophical paradigm. It is this paradigm, called "interactionism," elsewhere by Clark and Fast (2004), which is framed by the civic market theory.

Interactionism is, in short, the theory that because people (actors) interact in specific situations (everyday behavior such as business), companies and their behaviors are better understood. The decisions of business actors are not dependent upon numbers, figures, and statistics alone. Instead, business people form strategies and plans, such as deregulation public policy, knowing that they can maneuver the newly formed markets. A key component in understanding business in the interactionism paradigm is to also know, influence, or control the role of government. Much has been written on this subject, but the "invisible hand" of government needs to be influenced to do as business wants it to do.

Economists want to be scientific and therefore ignore this influence over government. Instead, they tend to think that the use of statistics and numbers place them above the interaction between people. Economists see themselves akin to the hard and natural sciences. There is almost a sense in economics that if the field is not scientific (e.g., statistical or numbers-oriented), then the field is not professional. For most economists, however, the perspective and view or definition of what science is and does, bears little proof in reality (Blumer, 1969).

Science is not a simple matter of statistics or numbers (Perkins, 1997). While some field or qualitative studies have been conducted, especially on productivity (see Blinder, 1998), economists remain steadfast in their belief that fieldwork is the main research area for sociologists and journalists. Yet, the need to explore the "productivity paradox" as Nobel Laureate, Robert Solow, called it in 1987, promoted statistical research in the 1990s only to crash land with the explosive truths behind the "productivity miracle" of that decade by the turn of the next century.

Clearly statistics did not tell the "truth" about productivity in the 1990s. The popular journal, The Economist (see Economist issues 1998–2002), often tries to "sugarcoat" or marginalize the accounting scandals of CEOs, major American corporations, corporate governance, and bankruptcies in 2002, as simply downward revisions of company financials, when in fact these crises represent only the beginning of corporate illegal misbehavior (Demirag et al., 2001). The issue of validation and verification of economic data is simply neither statistics and numbers (quantitative) versus fieldwork and observation (qualitative) data to prove points or hypothesis, but a combination of both (Casson, 1996).

Implementation of energy economics today has been traditionally done (prior to deregulation, privatization, or liberalization) through a variety of "mechanisms" by energy experts. HarvardWatch (2002) looked behind the scenes of public policy and discovered, however, questionable direct links between the objective experts at some universities and the energy private sector. The "links" between scholars and experts and the companies violates the credibility of economics as being either objective or scientific. Far more important are the "networks" of people who develop and implement government policies that impact the public through the private sector. As will be described below, government policies do not just mean regulations, tax, and incentive programs. They should also include, as California has championed, economic accounting for projects/programs (Schultz, 2001) and the creation of market demand (CCAA, 2001 and CAFCC, 2002, among others).

At this point, it is important to make note of how California government found itself in the middle of redefining energy economics. The energy crisis can never be fully explained (CEC, 2002) but one basic economic issue is clear: the state government had to take an active role in solving the crisis. For California, this meant a number of measures and legal steps had to be taken from long-term energy supply contracts to emergency funds for conservation and efficiency programs, to incentives such as buy down and rebate programs to expedited siting of new power plants.

As discussed earlier in some detail, other economists such as Borenstein et al. (2001), Woo (2001), and Nobel Laureates (2001) all agreed that the energy crisis could be averted and changed if the government simply took off all the price caps on energy. What is ignored traditionally by economists is a focus on the firm itself (Teece, 1996). Energy economics, however, only discusses the companies as end users of energy such that energy flow, hence costs, should be controlled by the consumer's awareness of "real-time prices" (Borenstein et al., 2001).

Elsewhere Clark (2004) argues that "qualitative economics" is a new area of economics, within the intereactionism economic paradigm. The purpose of qualitative economics is to understand how companies work. Much of field is concerned with case studies, corporate descriptions of operations, and people. But the most significant concern of qualitative economics is to gather data in order to understand what the meaning of numbers. For companies when they add, as Enron allegedly did, 2 plus 2 and got 5, the meaning of those numbers is critical. The issue is that economics must understand how businesses work and can only do that with deeper definitions, meanings, and backgrounds of organizations, people and their interactions.

The goal of economics must be to take quantitative and qualitative data and derive rules. Economics needs to expose universal rules based and tested in reality upon a combination of statistics and interactions. From rules, laws can bo articulated (Perkins, 1997). Scientists in fields such as linguistics (Chomsky, 1968, 1988) and developmental cognition (Cicourel, 1974), have long investigated science in terms of developing universal rules and laws. Just like the natural sciences, natural sciences use observation, description, and hypothesis testing (Chomsky, 1980). The science argument is spelled out in other works (Clark and Fast, 2004) and some aspects of the qualitative economic theories are presented below.

What is the role of government in business? Much of the economic literature has been focused on the dichotomy between free markets and tight regulation as in the historical electrical industry. The emerging era of public-private partnerships is neither. The justification for public involvement in the power industry is twofold. First the transmission and distribution monopoly and technological nature of electricity networks mean that the public has an interest in overseeing the private suppliers of such an essential part of modern life. This point has been made consistently in previous chapters. Second, the public has many social and environmental interests that intersect with the provision of electricity, such as environmental protection, public safety, equity, economic development, and long-term reliability. Simply put, given the extensive public agenda, it is more effective to try to reach these goals through partnerships than rule making. This is not unique to the electricity industry, though it stands out in very clear relief.

In Denmark, for example, the free market has historically involved in a partnership between government and business (Sorensen, 1994). If shared societal goals (free universal education, national healthcare, jobs, strong social services, and high standard of living) are to be achieved, then business and government must work together toward common economic goals. The "partnership" between government is not always smooth or cooperative, but it remains dedicated to the shared values for the common good (Sorensen, 1993).

Government is deeply involved in many industries in more than a regulatory role. For example, government provides over $16 billion annually to the US Department of Energy and its over dozen "national" laboratories. Two of these scientific laboratories receive over $1 billion annually in research funds: Los Alamos National Laboratory in New Mexico and Lawrence Livermore National Laboratory in California. Both of these labs as well as Lawrence Berkeley National Laboratory are operated by the University of California System which receives over $25 million annually as a management fee. The amount of research funds flowing through these and other laboratories clearly influences both public policy and business strategies in the United States and worldwide.

The recently passed national energy bill included assistance for coal and nuclear power as well as expanded incentives for oil and gas, as well as most parts of the electricity industry. The electronics industry credits high price defense contracts with giving them the capacity to develop and market early transistors and integrated circuits when there would have been no private markets for these products given their costs. In addition, the US agricultural incentives have become hotly contested by Europe and Asian countries claiming unfair competition in trade. Also, the Bush administration's favoring of government support for industry is seen in the prescription drug bill recently passed. In short, the myth of industry operating without government support and control is hopelessly inadequate.

The local and regional level is also a critical resource for public-private partnerships. The role of local governments is often forgotten, but together they have extensive planning and program activities because their residents and constituencies need and want it. Thus, local level governmental entities, such as government and counties, are one focal point for renewable energy generation and hence noncentral grid energy systems. In 2000, the voters of California passed Initiative 38 which allowed local governments or districts to use finance measures such as bond measures.

One of the most successful has been the Community College Districts, the largest college system in the world with 1.3 million students on 108 campuses. By the spring of 2002, six districts followed the lead of the Los Angeles Community College District (LACCD) and its Board of Directors who passed a bond for $1.3 billion. At least half of the bond measure funds are being used for "sustainable" (green) buildings in LACCD under international green building standards. In other words, the public colleges are leading the way to renewable energy in their facilities.

Part of the evidence of the political and economic success rests in the fact that the Board of Directors for the California Community College District appointed the Chancellor for the LACCD (Mark Drummond) to head the entire State System in 2004, the largest higher education systems in the United States with over 108 campuses and over 1.2 million students. The advancement of "green" college buildings throughout the state will certainly be far more rapid and cost-effective. The political and economic repercussions to this are significant. Local communities are the market drivers for renewable and clean energy systems.

In California, the mass purchasing of sustainable systems has reduced the price and expedited the implementation of these systems. The state and the agencies and programs it funds are a huge market and the mobilization of this market is large enough to change the economics for the production of many items. The state mandated some level of internal consumption of recycled paper, green building, and renewable energy, for example, which creates enough of a market to help establish these industries at a level where they have economies of scale and become cost competitive in open and unsubsidized markets.

Many examples exist of this financing practice such as the State purchase of police and other vehicles which are zero emission to the Energy Star program for energy-saving appliances and efficient equipment. The Department of General Services has led the state in this effort and included new technologies, such as fuel cells and solar devices (CSFCC, 2002).

Government partnerships rather than regulatory power help create more successful programs to meet civic goals. Using moral persuasion and the legitimacy of the state, governments can lead by demonstrating their willingness to invest in what they are telling private companies to do. California has the public policies and mechanisms in place for local and regional clean distributed energy systems. The State Government Code already provides for Community Energy Authorities (No. 5200) for local and regional energy systems (not municipal utilities). With local governments, the private sector, educational and research institutions, as well as nonprofits, all working together as partners, clean energy is viable along a business model for recovering costs and providing for innovation and change (Clark and Jensen, 2000). A number of these strategies include the following:

Promotion of dispute resolution and conflict mediation as a way to resolve differences between private firms and state agencies or programs. The competitive model is based on a win-loose model, whereas economic growth and public interests are increased with win-win responses to differences of opinion on what direction the power system should take.

The public role in partnerships is often and most importantly the collection of information and data on power demand, technological change, environmental resources and pollution, and national and global trends. As discussed in earlier chapters, one of the major contributions of the California

Energy Commission when it was formed was to help the utilities come up with a standard methodology forecast of power demand which led to the reduction in expected number of nuclear power plants needed. The fact these plants were not built protected the state from even worse disasters of stranded assets and high costs.

Resource mapping and technological feasibility studies are also an essential function that government can provide. Again, the Energy Commission had as one of its initial mandates the collection and dissemination of information on the location and extent of renewable resources such as geothermal hot spots, wind resources, untapped hydro capacity, solar capacity, etc.; these resource inventories became essential for firms looking to invest in these technologies in the state. These studies were done in partnership with the industry, pooling their private information and setting up methodologies for collecting additional data. Because the data came from a partnership involving the firms intending to use the data, it had a higher level of credibility than if it had been developed in order to force business partners to do something.

Provision of additional resources and technical assistance to implement best practices, including green building and site design, product development, and installation of energy efficiency systems (CAA, 2001). The public role in partnerships can also include developing and providing training, conferences, best practice reports, and consultations with staff experts that will eventually lead to more successful private developments meeting state standards and agendas.

Finally, the partnership can encourage California-based philanthropies and the commercial media to work with the public sector on public education and participation, and inform readers and viewers on energy issues. The partnerships in the state that encouraged conservation that eventually broke the price spiral of the energy crisis show that broad partnerships can work very effectively.

The implementation of new policies and programs cannot be done until basic public finance issues are addressed and resolved (Clark and Sowell, 2002). The state investment in conservation, new environmentally advantageous technologies, and public energy infrastructure is not an expenditure but a down payment that will generate considerable return. State Treasurer Phil Angelides had implemented two major policies that leverage the vast $300 billion State CALPERS retirement fund. Two policies were launched since 1999: (1) sustainable development as "Smart Investment" with a $12 billion investment and (2) the Double Bottom Line which is investing in California's Emerging Markets at $8 billion and another $1 billion to developing countries.

With globalization and increasing scale of corporate structures, the conditions for and effect of competition needs reassessment. Today more multinational companies have significant shares of the global and local markets, and they shape demand rather than respond to it. While large parts of the economy are strongly competitive, many trends of consolidation of firms, interlocking ownership and corporate-government linkages contrast with innovative and price competition characteristic of small business markets.

The new civic economic model must take into account the growing power of large firms and the inability of competitive markets to work effectively without oversight. It is the role of oversight that restrains these large firms from being uncompetitive and to exercise market power. In James Scott's (1998, p. 8) words, "Today, global capitalism is perhaps the most powerful force for homogenization, whereas the state may in some instances be the defender of local difference and variety."

Thus, the civic market is not built on the premise that a competitive market must be created and maintained; instead it is built on the premise that such a competitive market is impossible to guarantee and that the public good must be served and assured by active public partnerships between empowered state agencies and innovative and socially responsible companies.

Many economists have argued (see various issues in The Economist in February–March, 2002) that the collapse of Enron proves how neoclassic economics works (Clark, 2002). As the argument goes, if a company cannot perform in the market, then it fails. Other companies come in and take its place. The next economy moves on. The economics problem with that argument is that it ignores what allegedly Enron did in the first place—set up and influence the deregulation of the energy market; then manipulate it; and finally use or profit from investors so that in the end, the collapse of the company was based on inflated shareholder value and unsecured creditors who lost their funds. Employees lost their retirement, and with the collapse in the value of the stock, thousands of stockholders outside the company as well as mutual funds and retirement portfolios suffered. Thus, the consequence was not just a failed company; it was a cascade of pain for people all across the nation, and it was a leading contributor to the $40 billion electricity charge to the citizens of California.

However, there is a growing environmental economics tradition in which strategies for valuing the economy are made. For example, in business, making a profit may not be the only motive for the firm itself, individuals working within it, and shareholders themselves. One Danish researcher in the 1990s indicated that success is not defined totally by monetary reward. The survey results from 63% of about 350 researchers indicate that there are other rewards (what Clark calls deep structures in his books about "Qualitative Economics" 2008, 2014, 2019) for inventing and patenting new ideas and innovations. The Nordic countries have been leaders in their governments enacting policies and programs which protect the environment as well as create and establish policies, plans, and economic funding for both businesses and nonprofit groups to solve the problems of climate change—and even reverse it.

Take the case of Copenhagen, Denmark

Copenhagen started its work with climate change adaptation in 2008. The first step was the preparation of the climate change adaptation plan that was published in 2010 and approved by City Council in 2011. The adaptation plan looks at possible impacts on the city from expected climate change concerning rain, sea level rise, rising temperatures, etc. And at the time of the publication, climate change was still something for the future and it was expected that there would be a long process towards implementing the ideas of the adaptation plan.

But on July 2, 2011, a massive cloudburst hit the city. In 2 hours the city was flooded by 150 mm of rain—which in Denmark equals a 1000-year storm—a storm that statistically would only hit the city once every 1000 years.

The damages were massive; insurance companies received claims for more than 1 billion US dollars. But more importantly the cloudburst showed that the city was vulnerable. Critical infrastructure was heavily affected by the cloudburst with power outages at hospitals, closed roads, and threats to the city's emergency services.

The 2011 cloudburst became a game changer. All of a sudden climate change was not something that could happen sometime in the far future—it was here and now. And in the coming years more frequent cloudbursts underlined the need for action.

As a direct result, the city immediately started work on a cloudburst management plan.

The first step was to determine the future level of service based on our business case. Here we had three points:

- Just making cloudburst solutions was not a good business case.
- We needed to include normal stormwater management to make it pay off.
- Identifying the break-even point where repairing damages would be cheaper than preventing them. For Copenhagen, our analysis proved that break-even point to be a 100-year storm.

Once that was done we needed to look at methodologies. We had already in our adaptation plan done the basic economic calculations that showed us that an expansion of the existing combined sewer system would be too expensive and difficult to implement.

Climate change is not an absolute number—and neither is the city's development. There are so many uncertainties. We don't know exactly how severe climate change will be—and the city is also constantly changing. So we need to develop flexible solutions that can be adapted to the changing environment in the city.

Our methodology had to take these uncertainties into consideration. This is why we started to look at surface solutions. And not just surface solutions in isolated pieces dotted around in the cities. We are basically creating a new surface infrastructure to manage stormwater in the city. The system is city wide and connected—just like a normal sewer system, and it works parallel with the old sewer system keeping the stormwater out of the sewer system and in that way preventing sewer overflows and flooding of houses, etc.

A key element in the Copenhagen Cloudburst Management Plan is the added benefits of the adaptation work. By creating a surface infrastructure with cloudburst streets and retention areas, we also have the opportunity to create a better city. We can add more green, more urban nature, more recreational space, and add to the livability of the city.

After all, it does not rain all the time—so we have to make sure that our cloudburst interventions can function in the city when it is dry! From the very beginning of the adaptation work in Copenhagen, this has been a guide line for us—not just to solve problems, but also to look for opportunities.

In order to develop the Cloudburst Management Plan, the city was divided into seven water catchment areas. These catchment areas form the basis of the plan, and for each catchment area, hydraulic analyses were paired with analyses of urban landscape, topography, and urban development plans. On this foundation, a detailed plan for managing a 100-year storm was developed for each catchment area.

These plans were approved politically after a period of public consultation, and on the basis of that we started preparing the implementation plan. How do you actually plan how to implement 300 projects over a period of 20 years? Where do you start and where do you end—and what goes on in the middle? There were a lot of questions in the process. We needed to recalculate the economy, and we also had to make sure that the figures that we would be presenting for the final political decision actually were as accurate as possible.

In the implementation plan, we described all 300 projects—identified where we could see potential for greening, and synergies with other planning in the city. We dedicated a large part of our planning process to the economy and financing.

How does financing the Cloudburst Management measure work in Denmark? Basically, the financing comes from three different channels:

- Everything that can be clearly defined as stormwater infrastructure (pipes, tunnels, canals), with only that function, is financed and constructed by the municipality owned utility.
- Stormwater infrastructure combined with other functions (cloudburst streets, retention in parks, etc.) is financed by the utility but constructed by the city.
- Added urban space improvement on top of the stormwater infrastructure is financed by the city (taxpayer's money).

In the implementation plan, the total cost of the Cloudburst Management Plan was estimated to 1.5 billion US dollars. And in 2015, the City Council unanimously approved the implementation plan—and we were all set to go.

Now 3 years down the line, we have to admit that there are a few bumps along the way that is making the implementation more complicated than originally estimated. We need to make sure that the hydraulic modeling is more detailed and are currently in the process of revising the more than 50 cloudburst branches that constitute the plan.

Another experience is that it is not easy to implement a new infrastructure in a city. One of the obstacles is the different planning logics that we have. It is difficult for the utility's sewer engineers to work above ground—and there are issues like water quality, street design, traffic, etc. to take into consideration.

It takes time to reach a mutual understanding of how this has to be done. We are sure that it is possible. When Copenhagen started working with its bicycle infrastructure, we had the same problems—and 30 years down the line this is not the same problem that it originally was—everyone agrees that we have to have bicycle infrastructure everywhere.

What cities can learn from Copenhagen

We think that other cities can learn a lot from the Copenhagen case—both from our successes—and from our mistakes.

Probably the most significant thing to look at in Copenhagen is the business case approach rather than the focus on risk. Our success is that we were able to demonstrate that it would be a good idea to deal with cloudburst management in an integrated way, and that we could also work to improve the city.

The focus on the cobenefits of adaptation has proved to be important. Demonstrating that adaptation does not need to be only about preventing disaster, but can also improve the livability of the city by creating new recreational spaces, greening the city, and making it a nicer place also when it is not raining has been a key factor in the popularity of the cloudburst management plan.

A key message to other cities is to look at the city's need for development and link that with the adaptation work. This can also help leverage financing. If a developer is developing an area for housing, it is relatively easy to make sure that adaptation is a part of the development.

Another key message is that it is important to work with flexible and adjustable solutions. Climate change projections are full of uncertainties. We know that climate is changing, but we are not fully aware of how much and how fast. And cities are also changing. It is really difficult to fully imagine what the city will look like in 40–50 years' time. Solutions that can be adjusted are important.

A lot of cities have over time asked how much the cloudburst of 2011 has impacted the speed of the development of the cloudburst management plan, and is it necessary to have a disaster to get political backup? Other cities have looked at the Copenhagen case and used it to model what would happen if it happened in their city and used that to start the planning process.

The most important messages from Copenhagen would be:

- Make a plan and do the hydraulic modeling. Sometimes the solutions are far from where there is a problem.
- Link the adaptation work to the development of the city, mainstreaming it from the beginning.

In short—it is time to start working for changes today—not in a decade from now.

The chart below shows how there is a need for interactionism between (and within) governments at all levels and the public, businesses, and groups. As Clark and Fast (2008, 2019) argue in their books, the traditional economics theory of Adam Smith for "supply-demand" with the "government as an invisible hand" has never existed or even done. Instead, governments at all levels need to be part of the community planning, setting of goals that are then measured.

Case in point is China, aside from the international politics. The alignment of international business interests and China's "green development" policies were clearly visible following the Group of 20 nations (G20) Summit in Hangzhou, China, on September 3–5, 2016, where these global leaders pledged support of both the United Nations 2030 Agenda for Sustainable Development and the December 2015 Paris Accord for dealing with greenhouse gases, emissions mitigation, adaptation, and finance. The G20 is the premier forum for international economic cooperation.

The G20 support began in April 2016 when a group of personal representatives of the G20 leaders, known as the Sherpas, drafted the first Presidency Statement on Climate Change. That statement focused on how the G20 could take the lead in promoting the implementation of the Paris Accord and quickly bring it into force. The statement was seen by the Sherpas as "a significant step to advance international cooperation on climate change by sending a strong signal that the G20 is committed to vigorously implementing the Paris Agreement."

The second key to G20 support was advice from the Business 20 (B20) which held meetings in the 2 days prior to the G20 meetings. The B20 provides a significant platform for the international business community to support the work of the G20 by hosting focused policy discussions and developing recommendations geared toward strong, sustainable, and balanced growth in the global economy.

Mr. XI Jinping, President of the People's Republic of China, made the keynote speech at the B20 Opening Ceremony on September 2. Titled "A New Starting Point for China's Development, A New Blueprint for Global Growth," President Xi emphasized strongly the need for promoting "green development" to achieve better economic performance while mitigating and adapting to climate change (a concept reflected in China's new Five Year Plan, 13, adopted in March 2016).

"Green development" is the Chinese terminology for what the European Union calls the circular economy, whereby products, goods, and services are all seen as being reused and recycled to cut down on transportation, use less fossil fuels, and minimize waste all to reduce greenhouse gas emissions.

"Green mountains and clear water are as good as mountains of gold and silver," Mr. Xi told the B20. "To protect the environment is to protect productivity, and to improve the environment is to boost productivity. This simple fact is increasingly recognized by people." He went on to tell the B20 that "we will make China a beautiful country with blue sky, green vegetation, and clear rivers, so that the people will enjoy life in a livable environment and the ecological benefits created by economic development." Mr. Xi said this would be achieved through the reestablishment of the historic "Silk Road" for global trade, which was important long ago to China's development and cultural interactions with the West. Those trade corridors will now be expanded from the historic land routes between China and the Mediterranean to include oceanic routes today. "There is a need for action, not just talk," he said.

The conclusions the B20 forwarded to the G20 were titled "Toward an Innovative, Invigorated, Interconnected and Inclusive World Economy," and are included in the recommendations by the G20 in its Executive Summary.

The B20/G20 meetings resulted in the G20 Action Plan on the 2030 Agenda for Sustainable Development that recognizes the global importance of stopping climate change. From the Action Plan: "G20 efforts will continue to promote strong, sustainable, and balanced growth; protect the planet from degradation; and further cooperate with low income and developing countries. G20 members will ensure that their collective efforts make positive global impact toward effective implementation of the 2030 Agenda in all three dimensions of sustainable development in a balanced and integrated manner."

The G20 Action Plan is guided by fourteen principles and a series of collective actions that include renewable energy, technologies, climate, and "green" finance. Even though China continues to ramp up coal production, the nation still plans to meet its long-term obligations to address climate change and has criticized US President Elect Trump's pledge to abandon the Paris Accord. In a November 1, 2016, Reuters story, Xie Zhenhua, China's climate chief, said this about the United States: "If they resist this trend, I don't think they'll win the support of their people, and their country's economic and social progress will also be affected. I believe a wise political leader should take policy stances that conform with global trends."

That the B20 and the G20, along with China, recognized the importance of the Sustainable Development Agenda and the Paris Accord has encouraged support by countries around the world and accelerated ratification of the Paris agreement which required 55% of the 197 countries who assented to the agreement to ratify it. The 87 countries needed for implementation formally signed and approved the agreement in time for the accord to be entered into force by the November 4, 2016, deadline. As of November 30, 2016, 114 countries had ratified the agreement, including all G20 nations.

The long-awaited international Paris Accord is clearly an important milestone in the climate battle. However, as reported in the online publication EcoWatch on November, 29, 2016, the official policy of President Elect Trump will be climate denial when he takes office in January 2017. This leaves the future leadership role of the United States in question, as will global progress on climate.

Overview of city and community changes

Studies are now confirming how economics needs to become a science which needs to be both quantitative and qualitative data. The selection of strategies implemented in any particular city will depend upon the specific transpiration needs of the area with a goal achieving a high level of customer satisfaction across multiple rider types, including choice riders and transit dependent riders to link them with other ways to get around.

When developing first/last mile solutions, cities need to coordinate with transportation authorities to identify key locations for transit connectivity. Cities must also create cost estimates by strategy and location, implementation timelines, program governance and oversight, fare payment, and reservation technologies. This includes developing new ticketing systems or fares for riders that are connecting to multiple forms of transportation (e.g., using a bikeshare to ride to a train station). Additional actions to optimize transit access and increase multimodal usage include developing infrastructure improvements, changing in transit service levels, and creating public-private partnerships to support multimodal transit options from a wide range of service operators.

Japan has again stepped out to lead the world in applying technologies not only to cars and media but also to communities. Here is what Japan is doing due to climate change with more details and the focus on the City of Tokyo in Chapter 13 from Professor Andrew DeWit, School of Economic Policy Studies, Rikkyo University, Tokyo, Japan.

And there is more as other chapters and cases demonstrate with specific communities.

The impact on communities, neighbors, and individuals, yet the need for LED blubs and HVAC along with on-site power from renewable ("green") energy (wind, solar, geothermal, etc.) and "smart" systems are critical today in all buildings and their surrounding communities. The impact on real estate is deferent in states like California with its fires, rain, and droughts that have devasted real estate, land ownership, development and insurance costs as well as over 100 people have died. There needs to be solutions from governments at all levels with technologies, green energy, smart communities, and economics all tied together.

Case in point is the EU in this area now (2019) since 2015 when European Commission started working on Circular Economics (CE).

Completed and made public policy in the EU: Brussels, January 16, 2018. © European Union, 2020.

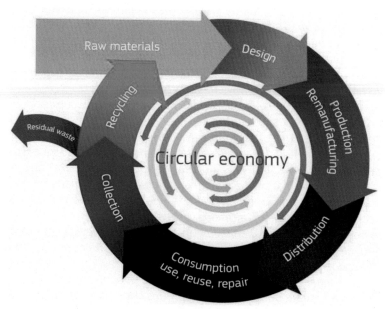

Commission staff working document: report on critical raw materials and the circular economy.

Reproduced from EU Commission on Circular Economics Brussels, January 16, 2018.

Aside from changing of the climate, the costs with new technologies and CE are the key to the solutions for smart, green, and healthy communities. Starting with technologies such as Clark's book on The Green Industrial Revolution *(2015), details are regenerative braking for transportation, energy storage, as well as solar panels for buildings and even thin film solar for cars.*

Now there are all-solar-powered cars as well as hydrogen fuel cell cars that use no fossil fuel and only electrized power from the sun. Along with electric vehicles (EVs) such as cars, buses, trains, and next airplanes for traffic flow due to travel, transportation and wireless, communication media, CE is key as it takes into account how products, systems, and technologies can be reused with externalities, groups, government, and global partners to set standards and measurements for buildings such as LEED standards, etc.

The chart shown below is what the Ellen MacArthur Foundation has been working for the last few years as it works with many groups in the European Union and now around the world including China and soon the United States.

The goals and results have been established using LEED standards (see below in Chapter 3) but can be summarized in the context of five factors due to the global population set to reach 9.7 billion by 2050, two-thirds of which will be living in cities, according to United Nations projections, governments worldwide are already facing an array of challenges in ensuring the sustainability, security, and success of an ever-more-crowded planet.

1. **Privacy fears**
    Amid fierce debates about how tech giants have made use of data, smart cities have become yet another flash point in the war over digital privacy. Even among those who support smart cities, there are fears that in practice, smart city technology will give Big Brother—be it big government or big tech—too much visibility into citizens' lives.
2. **Incomplete data**
    Even as privacy advocates voice concerns about too much data in the public sphere, there is also a data shortage that demands attention.
3. **Faster connections in a 5G world**
    Boasting fast connection speeds with ultralow latency, 5G will drive smart city development. For residents of Houston, Indianapolis, Los Angeles, and Sacramento, commercial 5G service is already a reality, and more municipalities are preparing to enter the 5G era themselves. Although the introduction of 5G will pave the way for a number of benefits, including better and faster communication between municipal systems and services, implementation will be no walk in the park and the difficulties are embodied by the providers such as AT&T coming under fire for "fake" 5G service.

4. Funding

   Given fiscal constraints, smart cities cannot be bankrolled by governments alone. Public-private partnerships, however, provide a pathway for funding smart city initiatives and enabling local innovation. Cisco, to take one example, announced in 2017 a $1 billion City Infrastructure Financing Acceleration Program, in partnership with a group of financial firms, to jump-start technological innovation in smart cities.

   As population growth strains infrastructure and places greater demands on local authorities, we will need more of such partnerships to play a vital role in supporting smart cities' growth. These public-private initiatives can also help overcome public resistance to the costs of major investments like high-speed rail.

5. Data storage

   Smart cities will run on big data, and that data will need to be stored: no cheap feat. As well as hacked and "borrowed" which is becoming a very common problem around the world.

   The concept and implementation of CE is critical to meet the needs of all communities including homes, work place, parks along with the energy, transportation, and Wi-Fi plus more current areas such as autonomous as well as all-solar-powered vehicles and the cloud for storage along other areas to come soon.

Megacities

What are megacities? As Wikipedia (2017) defines them:

A megacity is usually defined as a metropolitan area with a total population in excess of ten million people.[1] A megacity can be a single metropolitan area or two or more metropolitan areas that converge. The terms conurbation, metropolis, and metroplex are also applied to the latter. As of 2017, there are 47 megacities in existence. The largest of these are the metropolitan areas each having a population of over 30 million such as:

Asia clearly leads the world in terms of city populations over 20 million people.

Other Asian nations such as Delhi, India, has 27,200,000 people and Seoul, South Korea, has 25,600,000 people and Manila, Philippines, with 24,100,000 people, plus other cities throughout Asia.

China alone has 15 megacities such as Guangzhou (25,000,000 people) and Beijing (24,900,000) which will be topical areas later in book chapters. Europe has many examples too such as the Nordic countries have all been involved in sustainable cities for the last 3 decades. Case in point is Oslo, Norway (Europe's Eco Capital) as the world's smartest, greenest, and healthiest cities.

The Norwegian capital plans to cut emissions by 95% by 2030, despite being of the fastest growing cities in Norway and Nordic countries. Oslo was named the European Green Capital in 2019, and it hopes to be an example and role model for other cities.

A number of companies have been involved and worked with cities. One of the best is **Arup** Corporation (see www.arup.com and Martin Howell[1]) who developed communities using sustainable technologies, economics, and policy plans as noted below from many reports and sites.

Arup is an independent firm of designers, planners, engineers, consultants, and technical specialists offering a broad range of professional services. From 90 offices in 38 countries, our 12,000 employees deliver innovative projects across the world with creativity and passion.

Arup Foresight + Research + Innovation

Foresight + Research + Innovation is Arup's internal think-tank and consultancy which focuses on the future of the built environment and society at large. We help organizations understand trends, explore new ideas, and radically rethink the future of their businesses. We developed the concept of "foresight by design," which uses innovative design tools and techniques in order to bring new ideas to life, and to engage all stakeholders in meaningful conversations about change.

The built environment

The Report explores the meaning and applications of the circular economy within the built environment. It is intended to raise awareness of the circular approach, and to identify the many challenges, enablers, and opportunities available to Arup and others in making the circular economy a reality across the built environment.

As things stand, the application of the circular economy to this sector is less than straightforward. Existing frameworks eloquently express principles and philosophies. But they fail to offer specifics on how the built environment assets and services must be developed, procured, designed, constructed, operated, maintained, and repurposed.

The systemic nature of the circular economy requires both the ecosystem and its individual components to change. This means that governance, regulation, and business models could potentially be even more important to achieving the transition than design and engineering.

To get there, a dedicated built environment roadmap or framework is needed, together with a set of guiding principles for the design, engineering, and construction sectors. This will need to focus both on the economic

---

[1]Contact Martin Howell, CEM LEED AP, Associate Principal in Energy at Arup in Los Angeles, CA at Martin.Howell@arup.com.

business case and the opportunities to develop new ways to design and deliver projects. Such a framework would also help to drive innovation opportunities across the industry.

For further Arup information, details, and data, see:

Circular Economy: https://www.arup.com/perspectives/publications/ research/section/circular-economy-in-the-built-environment

- You can download our document "The Circular Economy in the Built Environment" from the link on this page.

Energy in Cities: https://www.arup.com/perspectives/publications/ promotional-materials/section/five-minute-guide-to-energy-in-cities? query=cities

- You can download the five-minute guide to energy in cities.

Future energy systems: https://www.arup.com/perspectives/publications/ research/section/energy-systems-a-view-from-2035

- Research showing likely future energy scenarios and technologies both in and out of cities.

Cities Alive: https://www.arup.com/perspectives/publications/research/ section/cities-alive-designing-for-urban-childhoods

- Focus is on Green Infrastructure and making healthy livable cities.

Smart Cities Strategies: https://www.arup.com/perspectives/publications/ research/section/smart-city-strategies-a-global-review

Public policy—the international perspective: overview

2

Woodrow W. Clark, II MA³ PhD

The world is round—so we need to think and go globally

Starting with the global policy level(s) related to sustainable megacities, China is clearly one of the leading nations to do that. While many cities in the EU and even a few in the United States have tried to create sustainable communities and cities, few have done it. In Chapters 1, 3—6, there are cases from around the world. And later in this Section 2, the Ellen MacArthur Foundation (from the UK) has a contribution to this area as they note that Circular Economics is global and are moving ahead with it in China since July 2018 when the EU and China signed an MOU.

Now next is the latest news (late May 2019) from the Chinese government—which has never been an "invisible hand" in economics. Frankly, no other government in the world has been an invisible hand either, but they will not say that as it is the traditional convention theory behind "western economics" (Clark and Fast, 2019). The Economist magazine had an "Advertisement" Report from the Chinese Government about the policies that China has enacted over the last few years since Chinese President XI Jinping took over.

The Report is titled "A New Stage: Green Development" which is about "The Belt and Road Initiative advances after years of hard work and development" authored by Yu Lintao (dingying@bjreview.com) as a result from the Second Belt and Road Forum for International Cooperation (BRF) April 25–27, 2019, where Chinese President XI proposed to build the Silk Road Economic Belt and the 21st-century Maritime Silk Road, collectively known as the Belt and Road Initiative in 2013. "The Belt and Road cooperation embraces the historical trend of economic globalization, responds to the call for improving the global governance system and meets people's longing for a better life," he added.

In its sixth year of development, more and more people have become familiar with the Belt and Road initiative and realize they need to engage with it, Kerry Brown, Director of the Lau China Institute at King's College, London, told Beijing Review. He noted that becoming part of the initiative conform to the interests of other participating countries and is a tangible way to engage with China, the second largest economy in the world. Though some countries still do not, they are in the minority.

At the Forum, over $65 billion cooperative agreements were signed. Now in May 2019, 126 countries and 129 international organizations have signed agreement with China. Italy and Luxembourg are the latest signatories.

Yang Jian, Deputy President of the Chinese think tank Shanghai Institutes for International Studies (SIIS), told the Beijing Review that he supported the Conference as "The reason why the Belt and Road Initiative is gaining wider and wider recognition is that besides promoting the development of participating countries, it has also contributed to the recovery of the world economy after the global financial crisis (Fall of 2008) and conforms to the UN Sustainable Development Goals."

Third-party market cooperation "plays a vital role" in "Belt and Road Initiative (BRI) which advocates and enables developed countries to play a vital role," Yang said. With the participation of more Western developed nations, "the misunderstanding over the initiative will be dispelled." Yang notes that "effective resource allocation and deep market integration among developed and developing countries in Belt and Road construction will create a win-win situation." Ueli Maurer, President of the Swiss Confederation, concerns with the BRI as it is a rare, long-term plan that has created a new dimension for the development of the world economy.

Over the past few years, BRI cooperation has expanded from the Eurasian continent to Africa, the Americas, and Oceania, "opening up new space for the world economy with better-than-expected results."

The Silk-Belt Road (SBR) needs to be "circular" (round) not "flat." The SBR needs to go east from China to Japan, Korea, Australia, and Southeast Asia. However, the SBR needs also to go to "nation-state" of California and the west coast of the United States as well as the entire nation of 50 states. The western hemisphere must be part of the SBR from Asia and also from the European Union (EU) where it started hundreds of years ago. The EU and the United States plus the entire western hemisphere need to be part of the SBR too.

China's development achievements have been shared with other Belt and Road participating countries as its huge demand for imports and increasing outbound investment has generated enormous growth opportunities. China's direct foreign investment in other countries participating in the SBR exceeded $80 billion. The total trade volume between China and those countries topped $6 trillion in 2013–2018, with over 244,000 jobs created.

Doing this will enact "Circular Economics" that is the other approach to economics (see Chapter 7) instead of the New-Classical Theory from Adam Smith which is "flat" not "round" as Clark's book Qualitative Economics (second edition) shows and his latest book (June 2019) Quantitative and Qualitative Economics: How to make Economics into a Science (Nova Press). President XI announced plans for the SBR to be a "meticulous painting" of the Institute through "extensive consultation, joint contributions, and shared benefits."

Since early 2018, the Ellen MacArthur Foundation (Cities in the Circular Economy: An Initial Exploration, 2018, pp. 1 and 10–11) has provided global leadership in public policy and government as a good place to start for "enablers" of urban areas to take actions as they are not an "invisible hand."

Circular economics

At the heart of creativity, innovation, and growth, cities play a central role as motors of the global economy. 54% of the world's population live in urban areas, and cities account for 85% of global GDP generation.[1] Cities are also aggregators of materials and nutrients, accounting for 75% of natural resource consumption, 50% of global waste production, and 60%–80% of greenhouse gas emissions.[2]

In the coming decades, cities will be increasingly important as even greater rates of urbanization are expected, and significant infrastructure investments and developments will be made. Cities could be uniquely positioned to drive a global transition towards a circular economy, with their high concentration of resources, capital, data, and talent over a small geographic territory, and could greatly benefit from the outcomes of such a transition.

As part of the Ellen MacArthur Foundation's ongoing research on the circular economy, this paper expands our understanding of the model in the urban context. This paper outlines some of the challenges cities are facing in today's linear economy, explores the alternative of a "circular city," and collates our research to date on the benefits of a circular economy for cities. Finally, it outlines outstanding questions on the topic, suggesting possible avenues of research for the future.

Urban transition to circular economics

Cities will play a substantial role in a global transition to a circular economy. There are a number of factors that uniquely position cities to drive the global transition towards the circular economy and greatly benefit from the outcomes of such a transition:

Proximity of people and materials in the urban environment
One of the main characteristics of cities is a high concentration of resources, capital, data, and talent over a small geographic territory. This proximity can enable the circular economy in a number of ways. Reverse logistics and material collection cycles could be more efficient due to the geographical proximity of users and producers, creating more opportunities for reuse and collection-based business models. The proximity and concentration of people enable sharing and reuse models (where products or assets are used multiple times by different users, typically within a neighborhood or smaller geographic unit).

Sufficient scale for effective markets
New circular economy business models are more likely to emerge and succeed in the presence of both a large and varied supply of materials, and a high-potential market demand for the goods and services derived from them. Both conditions are most likely to be met in cities.

The ability of city governments to shape urban planning and policy
Local governments have a large and direct influence on urban planning, the design of mobility systems, urban infrastructure, local business development, municipal taxation, and the local job market. Therefore, local governments can play an active role in embedding the principles of the circular economy across all urban functions and policies. Experts estimate that on a global scale, 60% of the buildings that will exist in 2050 are yet to be built, and in emerging economies such as India, this figure reaches 70% by 2030.[18]

Since these investments will largely need to be made in cities, it presents a massive opportunity for local governments to use their influence to apply circular economy principles from the outset for this infrastructure, which will help to avoid the "linear lock-in" currently seen in developed markets (where there is a need to transform large parts of existing, entrenched infrastructure systems). Local governments can use demonstration and pilot projects at the local level as showcases in order to engage national and business actors in the process as well. Where there's a lack of leadership on the national or regional level, local governments have been seen to step in. In addition, research suggests that local leadership is more trusted by citizens, indicating a stronger level of possible influence.[19]

Digital revolution

Digital technology has enabled a fundamental shift in the way the economy functions and has the power to support the transition to a circular economy by radically increasing virtualization, dematerialization, transparency on product use and material flows, and feedback-driven intelligence. Through the collection and analysis of data on materials, people, and external conditions, digital technology has the potential to identify the challenges of material flows in cities, outline the key areas of structural waste, and inform more effective decision-making on how to address these challenges and provide systemic solutions. Technologies like asset tagging, geospatial information, big data management, and widespread connectivity have been identified as enablers of circular economy activity in cities (pp. 10–11).

There is more but first is the need to define what sustainable smart, green, and healthy cities are around the world with some basic issues, policies, and plans. In the last 20 years, the private sector has invested heavily in the culture of the workplace, but the public sector has failed to keep up. As a result, governments face a growing "culture gap" which will imperil their ability to replenish their talent. Organizational culture is sometimes misconstrued as merely the perks a company offers, reducing the complex concept to what makes a company cool. It goes far beyond that.

Based on the data and definitions from Wikipedia, the United States has two megacities metropolitan areas where one is on the East Coast:

Organizational culture is the set of beliefs, values, and ideas that are learned and shared. Though often unspoken and implicit, culture determines how things get done.

As Shagorika Ghosh and Alexander Shermansong wrote in Meeting of the Minds (May 2019) on the Talent Crisis in the public sector with New York City is what Clark has done in most of his work to study and use of anthropology methodologies and theories. Ghosh and Shermansong take the same perspectives and strategies to analyze modern cities and communities, not just ancient and historical ones.

"Culture tells us what to do when the CEO isn't in the room, which is of course most of the time," per a now classic Harvard Business Review article. The most successful organizations align their culture with their business strategy, intentionally adjusting the culture to enable the strategy. When you overlook culture, your strategic aspirations are often stunted. As they say,

culture eats strategy for breakfast. City governments simply will not be able to implement 21st century policies with a 20th-century work culture.
   So how can city and state governments close the culture gap?

Assessing the culture gap: cross-sector partnerships

Surprisingly, there are few resources for addressing public sector culture, and sometimes government leaders dismiss the notion as too fuzzy. So let us suggest a very practical starting point: turn to your existing cross-sector partnerships. These partnerships and projects have been proliferating in cities across the country. Increasingly, cities are embracing the model of networked government in which municipal employees are not necessarily the direct provider of services, but rather enablers of those services.
   In addition to the direct value of the service provided, these partnerships offer cities powerful benefits such as:

- Exposure to new ideas. Through NYCx the CTO of New York City introduces municipal managers to emerging technologies and entrepreneurs that relate to the agency's mission and core priorities, from promoting the adoption of electric vehicles to reducing household waste to protecting small business from cyber risks.
- Requisite talent in the short term. The Startup in Residence program embeds startups within government agencies to codevelop solutions—bringing technology expertise that the agencies would not otherwise attract.
- Organizational learning and development. Pro bono programs like Civic Consulting Alliance in Chicago and the Silicon Valley Talent Partnership enable cities to borrow employees from local companies for special projects. These "professionals on loan" often bring new tools and techniques that can change how an agency operates even beyond the scope of the project.

Through cross-sector partnerships like these, city governments can begin to understand the culture gap. Increasingly, companies are using these very programs to enhance their own corporate culture.

Transforming culture: skilled volunteering

Many companies have already woken up to the fact that cross-sector partnerships can unlock tremendous HR value. As corporate surveys are increasingly showing, volunteering outside of your work increases employee engagement. In fact, skilled volunteering is one of the most cost-effective ways of advancing key HR objectives of attracting, developing, and retaining talent.

Here are three examples of how these partnerships can effectively address key levers in shaping culture within an organization:

1. Articulating core organizational values
2. Demonstrating the culture through highly symbolic actions
3. Continual renewal of skills and perspectives

The anthropology theory behind pro bono as a culture driver

Anthropologists have developed a concept of "liminality": status or privilege can change when individuals step out of their current status into a liminal state (limina is Latin for threshold). In the liminal state, they undergo changes, and then come back into a new normal. Similarly, when employees are given pro bono assignments, they step out of their regular milieu, often going to work at a nonprofit or city agency. Crossing that "threshold" gives them room to try new things, reflect, and grow. When they come back to business as usual, they are often changed by the experience—both refreshed and reeducated.
Intentionally selecting volunteer and partnership opportunities, executives can use "liminality" to reshape the workforce and the organizational culture.

Culture lever 1: articulating core organizational values
Too often value statements are merely posted on the wall, not actually embedded in processes. Skilled volunteering can change that. At Allstate, lIfe-time loyalty is an important value, as is employee initiative. The Allstate Fellows program reflects both values: employees with significant tenure can propose a community-oriented organization for them to help that fits both their specific skill sets and their individual interests. If accepted, the employee is sponsored to work for three months at that organization. This is a powerful incentive both to engage in the volunteering activity and also to build loyalty to the company.

Culture lever 2: demonstrating the culture through highly symbolic actions
Culture is also furthered through symbolic actions that express the culture and offer proof points. Take the case of Southwest Airlines, an organization that prides itself on connecting people to what is important in their lives. Southwest demonstrates this commitment through their Tickets for Time program—when employees devote more than 40 hours of volunteering to a nonprofit or school of their choice, Southwest Airlines provides that organization with complimentary roundtrip tickets, thus going above and beyond customer expectations.
The Salesforce model of corporate giving also leverages this model. In this case, when an employee finishes seven full days of volunteering (fully paid!). Salesforce donates an additional $1000 to a nonprofit of the employee's choice.

Culture lever 3: continual renewal of skills and perspectives

Volunteer projects allow employees to take on roles and responsibilities outside of their mandated tasks, which in turn can change their profiles and relationships within the organization. For instance, VMware, a company that values giving back to the community and their employees, has launched its Good Gigs program to promote service learning. They match employees with nonprofits that require specific skills, after which the employees must share their learnings from the experience with the rest of the company.

Putting it all together

When Investopedia launched their skilled volunteering program, they designed it holistically to engage employees' hearts and minds. The senior management chose to invest in financial literacy programs. After researching the field and reviewing hundreds of nonprofits, they nominated five organizations for pro bono time. Employees on loan helped these organizations align with Investopedia's mission and pitch themselves to the entire company.

All employees then voted, from the interns up to the CEO, collectively selecting two nonprofits to help them volunteer in high schools and teach financial literacy. This program demonstrated Investopedia's value of democratizing financial literacy, honed employees' analysis and presentation skills, deepening financial knowledge, and analytical capacities; while also helping drive the brand and increase the number of people visiting the Investopedia website, thereby enhancing revenue.

Applying these models in the public sector

Closing the culture gap begins with two steps.

1. City agencies can partner with a civic-minded company to observe and learn how they treat culture as an organizational asset.
2. Agencies can create their own skilled volunteering program to reshape the culture.

It might seem unrealistic in the public sector, where budgets are already stretched so thin, but there are reasons to hope. For example, the US Department of Education encourages employees to participate in tutoring activities and volunteering at schools. And Strong Cities, Strong Communities created a program to loan federal employees to be embedded in mayoral offices to promote the implementation of federal mandates.

In fact, research has shown that public sector employees volunteer at higher rates than their for-profit counterparts. If elected officials and public agencies harnessed this public service motivation, it would be a tremendous force to enhance their organizational culture, even without additional investment.

As cities surf the silver tsunami, cross-sector partnerships can provide two critical benefits. In the short term, the partnerships provide critical talent for policy priorities. Longer term, municipal governments can learn and emulate how private sector leaders have shaped organizational culture.

As we face continued challenges and a faster pace of innovation, the cities that make their work cultures attractive to high-potential younger employers will continue to grow and thrive.

Sustainable smart, green, healthy cities and communities

By Woodrow W. Clark II, PhD (*)

The world has changed. Other cities in Asia as well as in the western hemisphere, Africa, Asia, and a few in the EU have over 10 million people in them. The big issue again is "qualitative." That is how megacities are defined. Then measured, reported, and evaluated as to their being "sustainable." The book gets into these details with qualitative definitions and information as well as quantitative in terms of tracking statistics, numbers, and results.

The United States needs to follow as well to "leapfrog" other nations as it has done historically. Other nations, especially in Asia and the Nordic countries, have enacted ideas and programs that green their communities as well as being both small and healthy. Hence there are more and more opportunities, especially in the real estate "market" which is no longer just one building but many that are linked, circular, and collaborative with each other: energy, water, living, transportation, waste, etc. All of this along with "green" yards, roofs, interior design, and systems as well as transportation being not only nonfossil fuel but using on bikes, scooters, and designated areas to move around communities.

Case in point are also companies and their becoming sustainable by taking actions which are defined and measured. Google is one in Silicon Valley, California, USA, where "Climate Action in partnership with the UN Environment Department" reported in December 2016 that "Google (Headquarters in Silicon Valley) would be "Powered by 100% renewables" by 2017. The title of the report was "Internet and tech giant Google has announced that all its data centers and offices will be powered fully by renewable energy from next year.

So Google was the first corporate buyer of renewable electricity in the world with 44% of its electricity bought from solar and wind plants in 2015. The company has now committed to buying 100% of its electricity from renewable sources from next year onwards. Yet the 100% renewables target does not mean that Google will obtain all its energy directly from wind and solar every year. The total purchases from renewable energy equal the electricity needed for its operations.

Marc Oman, EU Energy Lead at Google, said: "We are convinced this is good for business, this is not about greenwashing. This is about locking in prices for us in the long term. Increasingly, renewable energy is the lowest cost option… Our founders are convinced climate change is a real, immediate threat, so we have to do our part."

Tech companies currently represent 2% of global greenhouse gas emissions, which is almost as high as the aviation industry's emissions. The company offices host 60,000 staff, and the data centers require an increasing amount of power, despite the improved energy efficiency brought by the use of AI. With a growing concern about climate change and global warming, tech companies have come under increasing pressure to reduce their carbon footprints.

Google originally set its target of being powered by 100% renewable energy in 2012, and it took them 5 years to achieve this goal—due to the complexity of power purchase agreements negotiations, according to Oman who said: "It's complicated, it's not for everyone: smaller companies will struggle with the documents. We are buying power in a lot of different jurisdictions, so you can't just copy and paste agreements."

The Internet giant bought 5.7 TWh of renewable electricity last year—mostly from wind farms in the United States, the equivalent of more than two-thirds of the whole of electricity generated by solar panels in the UK in 2015.

Google is also looking at nuclear, hydro, and biomass which is a big mistake if using nuclear power and reverses their goals due to the waste from nuclear power being a massive environmental problem as what to do with it!

Oman said: "We want to do contracts with forms of renewable power that are more baseload-like, so low-impact hydro; it could be biomass if the fuel source is sustainable, it could be nuclear, God forbid, we're not averse. We're looking at all forms of low-carbon generation." He added: "We don't want to rule out signing a nuclear contract if it meets our goals of low price, safety, additionality and in a sufficiently close grid, we don't want to rule that out, but today we can't positively say there are nuclear projects out there that meet this criteria."

Starting with the "smart city" are the methods for getting the information which is both qualitative and quantitative. Thus history, background, and key sources are needed.

Government: plans, goals, and strategies to be smart and healthy communities

Woodrow W. Clark II, MA³, PhD

Founder/Managing Director, Clark Strategic Partners, Beverly Hills,
CA, United States

Getting started

Regardless of where cities are in their journey, smart city and service planners must get ahead of the "trust curve" that is taking steps to build out a trustworthy ecosystem. Smart city initiatives are top of mind among city leaders, urban planners, and technology vendors today. According to IDC (International Data Corporation), these initiatives will translate to $80 billion of technology investments worldwide in 2018 and $135 billion by 2021. Although the smart city may be powered by technology and data, it is enabled and sustained by the trust its stakeholders have in it.

A smart city works when there is an implicit trust in the relationship its stakeholders have with the city, its services, its service providers, and with each other. When there is no trust or that trust is lost, its stakeholders will not use its services, nor will they be able to collaborate effectively with each other.

Even worse, they will stop participating and providing critical information. Its services will be underutilized as people seek alternatives. Smart city services will lose its value, as expected outcomes do not materialize because they lack the critical input and support from the stakeholders it was intended to serve.

To be relevant and remain so, a well-functioning and sustainable smart city must design trust in right from the outset, and continually reaffirm it. This trust must be integrated into all aspects of the smart city—its people, processes, management, policies, and technology. Case in point is the environment and historical aging issues of a city.

Smart city initiatives are at the top of the minds among city leaders, urban planners, and technology vendors today. According to IDC, these initiatives will translate to $80 billion of technology investments worldwide in 2018 and $135 billion by 2021. Although the smart city may be powered by technology and data, it is enabled and sustained by the trust its stakeholders have in it.

A smart city works when there is an implicit trust in the relationship its stakeholders have with the city, its services, its service providers, and with each other.

Sustainable Mega City Communities. https://doi.org/10.1016/B978-0-12-818793-7.00001-9

When there is no trust or that trust is lost, its stakeholders will not use its services, nor will they be able to collaborate effectively with each other. Even worse, they will stop participating and providing critical information. Its services will be underutilized as people seek alternatives. Smart city services will lose its value, as expected outcomes do not materialize because they lack the critical input and support from the stakeholders it was intended to serve.

To be relevant and remain so, a well-functioning and sustainable smart city must design trust in right from the outset, and continually reaffirm it. This trust must be integrated into all aspects of the smart city—its people, processes, management, policies, and technology.

Air transportation: an example of trust in action

The air transportation industry is built upon "trust." Very few people will willingly fly in a plane if it was not safe, nor will they use it to ship packages if it would not "absolutely positively" get there safely and on time. Fortunately, air transport is safe and reliable. According to the June 2018 International Air Transport Association (IATA) industry statistics, air carriers reported 4.1 billion passenger miles flown and 61.5 million freight tonnes shipped globally. Despite this volume, air transport is relatively safe. The IATA reported 45 accidents in 2017, down from 67 in 2016 and a high of 86 in 2013. Accident rates, per million miles, were 1.08 in 2017, down by half over the previous 4 year average of 2.01.

An air transport "trust ecosystem" ensures that flying is safe and reliable. A combination of rigorous engineering, regulations, policy, operational processes, stringent oversight, and maintenance has made air transport remarkably safe. An ecosystem of partners, from government agencies, aircraft and component manufacturers, airlines, engineers, and others have worked together to ensure these outcomes.

The air transportation industry is built upon "trust." Very few people will willingly fly in a plane if it was not safe, nor will they use it to ship packages if it would not "absolutely positively" get there safely and on time. Fortunately, air transport is safe and reliable. According to the June 2018 IATA industry statistics, air carriers reported 4.1 billion passenger miles flown and 61.5 million freight tonnes shipped globally. Despite this volume, air transport is relatively safe. The IATA reported 45 accidents in 2017, down from 67 in 2016, and a high of 86 in 2013. Accident rates, per million miles, were 1.08 in 2017, down by half over the previous 4 year average of 2.01.

An air transport "trust ecosystem" ensures that flying is safe and reliable. A combination of rigorous engineering, regulations, policy, operational processes, stringent oversight, and maintenance has made air transport remarkably safe. An ecosystem of partners, from government agencies, aircraft and component manufacturers, airlines, engineers, and others have worked together to ensure these outcomes.

Mention trust in any smart city conversation, and it quickly turns into a privacy and cybersecurity discussion. Although important and relevant, they are only two elements of many that create trust in tomorrow's smart cities. Distilling the concept of trust (and ultimately the solution) into these two elements oversimplifies the challenges involved and misses important underlying contributing factors. This results in inadequate approaches and solutions in creating trusted smart cities.

Although smart city planners speak in trust, residents and businesses are concerned with outcomes. When a city creates services that provide the outcomes residents and others expect and rely on, at a consistent and fair price, then the sense of trust is earned and reinforced. Residents expect that the bus service gets them to work and brings back home safely and on time every day. They expect police and emergency services will arrive quickly regardless of whether they live in a rich or poor neighborhood. They expect that redlight cameras are used to catch traffic violators, and not to track their movements around the city. The foundations of health buildings are noted earlier and later.

Trust defined

We define trust in the city as the firm belief in the outcomes and value of the services provided. These services may be provided by the city or others in the smart city ecosystem. These trusted outcomes are relevant, rendered reliably and with integrity, by service providers who are credible, transparent, and have the capacity to execute. Trust forms when a smart city consistently develops services and outcomes that are relevant when it is appropriate, of value to its users, and perceived to be reasonably priced. Any city or service provider that fails to do this, risks being called out for misusing taxpayer money and resources.

Services and outcomes must be rendered reliably, without bias, and with integrity. No one would use a service if its outcomes cannot be delivered reliably and consistently. No one would use ridesharing if they cannot find a ride and get to their destination on time safely. Equally important, services must be rendered to deliver the right outcomes they were designed for. They must be rendered accurately, legally, and in compliance with regulations or policies, and for its intended use and not any other use.

Service providers must be credible, transparent, and be able to deliver the desired outcomes. They must be knowledgeable and experienced in their field, trained, and if appropriate, certified. When rendering services, they must be transparent, fair and unbiased, collaborative and adaptive when necessary. Finally, they must have the capacity and capability to consistently deliver the services and outcomes.

Ten levers of trust enablement

To create trusted outcomes, planners and services providers have several "levers" that they can use. The careful selection, coordination, and application of these levers will build trust in new services, or increase trust in existing services. The levers include the following:

- People and Organization—skills, expertise, and certifications; jobs, roles, and responsibilities; teams and organizational structures for execution—often spanning functional silos.

- Governance—management, policies, legislation, audits, oversight, documentation, traceability, and metrics
- Processes—control and determine how services are developed and rendered
- Data—information used to build the service, as well as information created as a result of the service. Data must be representative and unbiased. Data must be protected and private.
- Algorithms—machine learning and other methodologies that define the service, as well as how it is delivered to its users
- Technology—tools for developing and operating the service, as well as connectivity, infrastructure, and solutions for delivery of the service
- Security—safety, privacy, and protection of assets, tools, infrastructure, and users
- User Engagement—how service providers and users interact with the service and its outcomes
- Strategic—vision, planning, partnerships, funding, and resources
- Change and transformation management, meaning the engagement of affected stakeholders before, during, and after, includes listening, communications, accommodating diverse views and perspectives, reporting performance to all stakeholders, training, and cultural management.

The ability of the city and its service providers to strategically develop and deploy these levers is a core competency and a critical success factor in building a smarter and more responsive city. Over time, as the service matures, and as users' expectations and needs evolve, these levers must be continuously examined and adjusted to maintain or exceed current levels of trust.

Putting it all together: the smart city trust framework

Trust does not magically happen by itself in a smart city. It must be carefully embedded and orchestrated across the entire smart city ecosystem. It must be incorporated into the processes and programs that create smart city services throughout its lifecycle, from design, to development, launch, operation, delivery, and use to ensure "trusted outcomes". Equally important, building trust is not the sole responsibility of the city itself, but everyone who "touches" the smart city—from beneficiaries, service providers to planners and policymakers.

The framework for building trust into the smart city is created when services from the various value creators in the smart city ecosystem create outcomes that are relevant, rendered reliably and with integrity, by service providers who are credible, transparent, and have the capacity to execute.

However, the elements of trust must be considered and incorporated at each phase of the services lifecycle. This is done by examining the various levers of trust, and incorporating the appropriate specific tactics at each lifecycle stage, to create the proper outcomes. There is no one size fits all set of tactics. Although there may be some common tactics, they will vary from service to service, provider to provider and city to city as well as state to region and nation to nation around the world.

Take the example of a city that wishes to deploy smart and connected streetlights. This new service allows the city to remotely monitor and manage its streetlights. If a light is not functioning properly, they will know immediately and schedule a repair. The connected lighting system also incorporates machine learning algorithms that will proactively alert the city of impending failures so that it can be serviced before an actual failure can occur.

Upon the deployment of this smart lighting service, business as usual is not good enough. Trust in this service and the public works department (and ultimately the city) is diminished when the lights are not repaired quickly because there is now no excuse for not knowing a streetlight is out. Trust in the service is diminished by the public works team if the technology indicates that the lights are functional when it is not, or if the light needs servicing when it does not.

Resident and other stakeholders must be engaged to understand their concerns, needs, and expectations. Current processes, performance metrics, policies, and systems must be redesigned to allow the city to respond to outages in days, not weeks. In addition, changes to the organization, including jobs, roles and responsibilities, team structure, culture, and accountability reporting will be necessary. To minimize the disruption caused by these changes and facilitate adoption by both city employees and residents, change and transformation management activities, including communications and training, must be conducted before the actual service changeover.

The new service is only as effective as the technology, algorithms, and data enabling it. Machine learning algorithms must interpret the sensor data from the remote streetlight controller units and distinguish between a true anomaly and a false positive. In addition, these algorithms must predict with high accuracy when the lights may need servicing by examining the sensor data. To maintain this accuracy, these algorithms must be continuously trained by certified experts using representative and unbiased datasets. Finally, the connected streetlight system must be secure, and not allow for unauthorized access to the lights nor disrupt the metering and billing systems.

Key next steps are as follows:

- Understand the trust ecosystem framework and adapt it to the realities of your specific city. The framework shown here is a template for reference purposes. Adapt and build upon it to suit your specific vision, strategy, and execution plans.
- Understand the current state of your trust capabilities and trust ecosystem. Conduct a high level "trust diagnostic" using the framework to quickly identify areas of strengths and weaknesses. Build a trust development plan.
- Evaluate ongoing and planned smart city services and projects against the framework. Use this framework to identify what is missing from the project plans and what is needed to make the projects more "trustworthy."
- Build your "trust capabilities" by identifying and focusing on high priority areas. Build your capability through a combination of internal development, outsourcing, augmentation, and strategic partnerships.

International policies that create programs
How smart districts can drive urban innovation

By Jonathan Adrews (13 December 2018. Managing Editor of *Cities Today*, PFD's flagship title on urban management and development. He focuses on data management and the application of IoT in cities)

Hosted by Dublin City Council in the Smart Docklands area, the value of smart districts was a key point of discussion during the 14th 20-20 Cities roundtable meeting, convened by *Cities Today*. The Smart Docklands district was officially launched in February this year. Jamie Cudden, Smart City Programme Manager, Dublin City Council and cochair of 20-20 Cities, said that what was lacking previously was a platform for start-ups, big tech, the city, and academia to come together.

Essentially we have a collaboration model where we as the city can act as the independent voice and broker," he said. "Cities need to be at the heart of that, to shape the application of emerging technologies in a way that benefits the city and its citizens.

Smart DocklandsCollege Dublin. Theo Blackwell, Chief Digital Officer, London said that involving universities was crucial to the success of a smart district.

Today smart districts are not only home to university campuses but also spaces where new and unique collaborations between university departments and cities can be developed. UCL's new space at Here East in the Olympic Park, London between the Bartlett Faculty of the Built Environment and the Faculty of Engineering Science is one such example and another is what we see through the Smart Docklands here in Dublin.

Nearly half of the 15 city chief innovation, technology, and digital officers in the room said that they had a smart district of their own—all at various stages of maturity. James Noakes, Councillor and Cabinet Member for Energy and Smart Cities at Liverpool City Council, asked his peers if such districts are better off being allowed to develop on their own."Some creative areas of cities have sprung up organically," he said. "They were created by people because the city knew when to step out of the way and only stepped in when they were called upon."

Similarly, how to recreate and foster such creative areas of a city needs to be better understood. "Why is this particular area booming?" asked Maddie Callis from City Possible, Mastercard. "People went there first and then it became smart; what is there to be learnt?"

Others asked if smart districts would still be relevant in 10 years' time, joining the likes of science parks. Bart Rousseau, Chief Data Officer, Ghent, said that his city lacks new land or brownfield sites to create a district of its own from scratch.

"We can look at hub areas if that is the case," he said. "Proximity [of hubs or smart districts] to the rest of the city is an asset, as is ensuring there is benefit

and value to people. Retrofitting a medieval city like mine with 5G will create noise, roadworks and complaints from residents. We need to bring this back to making a better life."

Ensuring social issues are included is equally important, noted Trevor Dorling, Director, Digital Greenwich, London. Michael Guerin, Programme Manager, Smart Docklands Dublin, said that being close to the rest of the city and in the middle of one of the largest social housing areas has meant Smart Docklands is a driver of community development for local residents and business.

Similarly, Paddy Flynn, Director of Geo Operations, Google said that as the largest corporate resident—the tech giant established its European headquarters in the Docklands 15 years ago—it has been heavily involved with the local start-up ecosystem and community development.

Other topics discussed over the two-day meeting included multicity accelerators, data sharing with mobility services, and digital rights and data commons. For more information on the 20-20 Cities meetings, see 2020cities.com.

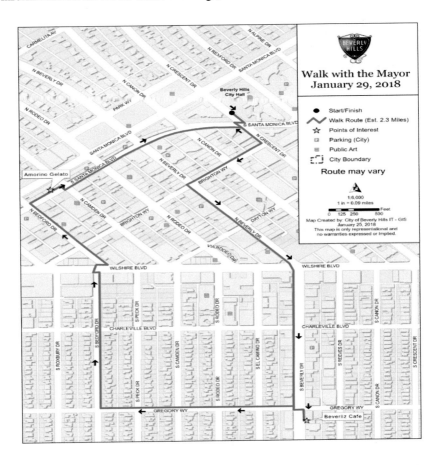

Nordic countries: the case of Stockholm, Sweden

Estimating consumption-based greenhouse gas emissions at the City Scale.

A guide for local governments.
SEI (Stockholm Environment Institute (February 2019)
Derik Broekhoff, Peter Erickson and Georgia Piggot

Introduction

Numerous cities around the world have been exploring their carbon footprint using consumption-based emissions inventories (CBEIs) (Millward-Hopkins et al., 2017; Erickson et al., 2012; BSI 2013; Jones and Kammen, 2015). These inventories differ from the territorial (or "sector-based") approach typically used to calculate urban GHG emissions because they include emissions generated outside city borders to produce goods and services for urban residents. These upstream emissions can be significant, and they are often commensurate with the amount of emissions created within the city itself (Pichler et al., 2017; C40 Cities, 2018). A CBEI can lead to insights about where local consumption gives rise to emissions outside a city's borders and suggest additional opportunities for reducing emissions.

This report provides an overview of what CBEIs are and how they can be used (i.e., the kinds of insights they can generate, as well as their limitations). It also describes the primary methods for constructing a CBEI, including "rough approximation" methods that can be used where more intensive methods are impractical.

In four sections, this memo states:

1) Briefly summarizes the basics of a CBEI
2) Reviews the key insights generated from CBEIs conducted in cities today
3) Describes the tools and approaches a city government could use for assembling a CBEI
4) Outlines considerations in choosing among different tools and approaches

An overarching consideration is whether and how the methods used to develop CBEIs inform the prioritization and development of local climate actions, including policies and other interventions that drive changes in consumer behavior.

The basics of consumption-based greenhouse gas emissions

A CBEI is a calculation of all of the GHG emissions associated with the production, transportation, use, and disposal of products and services consumed by a particular community or entity in a given time period (typically a year). In this way, a CBEI can create a comprehensive emission "footprint" of a community.

At the most basic level, the calculation of a CBEI entails two main inputs: a measure of what is consumed, multiplied by a measure of how many GHG emissions are associated with each unit of consumption. For example, if a community consumes one million tons of cement each year, and producing and transporting each ton of cement released 1 ton of CO_2, the community's CBEI would include 1 million tons CO_2 associated with cement consumption, regardless of whether that cement was produced inside or outside the community's jurisdictional boundaries. The same basic approach would apply for a unit of any other product or service, whether that be broccoli (there is about 1 ton CO_2 per ton of broccoli), beef (more like 20 tons CO_2 per ton of beef), or ballet tickets.

A full CBEI would, in principle, estimate the emissions for all of the products and services consumed in a city (or by a city's residents). However, this can prove challenging in practice because calculating consumption-related emissions can be highly complex. Communities consume thousands of different types of products and services, and the emissions associated with each of these are affected by many decisions made by different actors throughout their life cycles (e.g., production, transportation, distribution, use, and disposal). It would take a substantial amount of effort to track and understand the emissions associated with every unit of consumption to create an accurate CBEI—and any estimate is likely to become obsolete as production processes and supply chains change over time (sometimes month-to-month or week-to-week).

Fortunately, cities can now produce relatively simple estimates of CBEIs, thanks to recent innovations in the tools available. Furthermore, cities have completed enough CBEIs (including as part of a recent study by C40 Cities) to enable insights about what kinds of policy-relevant information can be distilled, even where methods are less precise or have limited resolution in certain consumption categories (C40 Cities, 2018).

The sections that follow provide some of these key insights, as well as further information for policy-makers interested in conducting CBEIs for their own communities.

Key CBEI insights

One of the first insights of a CBEI for most cities (at least those cities without substantial heavy industry within their borders) is that a full emission footprint can be substantially higher than a typical, territorial GHG emissions inventory. This is because the emissions associated with goods consumed in cities are often released outside the city during production and transportation, such as in the course of mining or growing raw materials, processing and manufacturing these materials into products, and packaging and delivering them to consumers.

For example, a CBEI for cities in North America (the United States and Canada) is typically about twice as high as "territorial" emissions, or emissions that occur primarily within a city's geographic boundaries; such emissions are associated with local transportation, buildings, power, industry (if any), and waste. The ratio of consumption-based to territorial emissions in European cities can be even larger, as territorial emissions in many European cities are even smaller than North

American cities (on a per-person basis). C40 Cities' recent study of consumption-based emissions applied a common methodology to 79 cities around the world and found similar results. For the sample of cities examined, CBEIs were 60% higher on average than territorial GHG inventories (C40 Cities, 2018). This average, however, masks considerable variation among individual cities. For some, mostly in Europe and North America, the ratio is much higher; for others, primarily in Asia and Africa, consumption-related emissions are lower than GHG emissions. Cities with higher territorial emissions are effectively net "exporters" of emissions, either because they produce more than they consume or per resident their production is more GHG-intensive than their consumption (or both).

Canada, Norway, the UK, and Sweden are all at least twice as high—and up to four times as high—as the global average. This implies that industrialized countries should take more responsibility for mitigating.

The second key insight from calculating CBEIs is that for cities in relatively wealthy, industrialized countries, the full footprint of residents' consumption is—unsurprisingly—far higher than the global resident average. Consumption-based emissions associated with residents in the United States, Canada, Norway, the UK, and Sweden are all at least twice as high—as the global average. This implies that industrialized countries should take more responsibility for mitigating global climate change, as contributions to and up to emissions are not distributed evenly.

A third main finding is that certain types of consumption consistently dominate a CBEI. Emission from car travel, building heating, and power consumption each average at least 2 tons CO_2e per person in the United States. But this is already well known from the standard, territorial GHG emissions inventories of cities. The unique contribution of a CBEI perspective is the expanded emphasis placed on goods, food, services, and air travel, each of which also is generally responsible for another 2 tons or more of CO_2e per person. (Depending on how new building and infrastructure construction are categorized, these too may contribute another 2 tons or more per person).

The economics of sustainable development

JudiGail Schweitzer-Martin, MRED, AMDP, CGBP, CALGreen CAC, ENV SP, SBE/DBE

Adjunct Professor, University of California, Irvine President — Chief Sustainability Advisor, Schweitzer + Associates, Inc., Lake Forest, CA, United States

"Applied sustainability: Optimizing value from concept, to the boardroom, to the [triple] bottom-line"

In successful, vibrant and appealing community developments, no matter their size, all the elements—housing, transportation, employment, commercial, recreation, education, services—must work well together and enhance each other. A thriving community is a system of interconnected, layered uses where weakness or lack in any element affects all the others.

Nature is also a system, or really a system of systems. Without human intervention, nature offers an elegant model of sustainability. Nothing is wasted. What decays here becomes food there; what evaporates there becomes precipitation here. "Sustainable" development means just that—a built environment that has internal logic, balance (taking from all sources in equal measure to what it returns), and the capacity to regenerate itself over time. If something is "sustainable," it is also, by definition, economically viable. Our health and wealth is interdependent on a healthy sustainable economy and healthy sustainable environment.

The terms "sustainable development" and "green" are sometimes used interchangeably when in fact they are not synonymous. The term *Sustainable Development* is a broader concept defined in the United Nations Bruntland Report (a.k.a. Our Common Future), as "Development which meets the needs of the present without compromising the ability of future generations to meet their own needs." This concept includes not only a single bottom line for developing our communities, rather a triple bottom line (TBL) of financial capital, environmental capital, and human (or social) capital. Sustainable development can be described as a process of developing land, cities, businesses, and communities that are economically viable as well as environmentally responsible and socially equitable.

Sustainable Mega City Communities. https://doi.org/10.1016/B978-0-12-818793-7.00002-0

The triple bottom line

The TBL attempts to measure sustainability by capturing the economic, environmental, and social impacts of an organization's activities on the world. In concept, optimizing the added value of sustainable design and development involves investing at the "sweet spot" of mutually beneficial aspects of financial capital, natural capital, and human capital.

Common practice of developers not accounting for environmental capital and human capital led to the development of a TBL accounting framework that could be used as a "lens" to "account" for not only market, or financial costs and benefits, but also for environmental and social costs and benefits. Previously, these quantitative and qualitative aspects have been considered "externalities." Total cost accounting recognizes these otherwise invisible costs and benefits and brings them into focus in decision-making, design, planning, and financial analyses.

Total cost pricing

As with any good business decision, accounting for direct and indirect financial costs (first time costs, operational and potential long-term liability) and direct and indirect benefits (being able to sell your product, being able to sell it at a premium, being able to sell your product faster than the competition, and developing a strong brand) are as important as understanding potential "opportunity costs". This is true for sustainable development "accounting"; however conventional accounting is expanded to include both "market" and "external" costs and benefits ("use value"); with the additional value of "opportunity benefits" of "nonuse value" (not consuming what would have otherwise been consumed in the business as usual "base case"). Conventional financial measures.....(ROI) are important, however there is real value in "avoided costs" and not having to build additional power plants for the energy we would have consumed compared to a business as usual base case; or reducing our dependence on nondomestic nonrenewable energy resources has a value and it can and should be counted. Accounting for both positive and negative "externalities" requires unique accounting tools that address total costs and benefits.

Total cost pricing (TCP) combines aspects of all of the above analytic approaches into a single presentation, such as the Total Cost Pricing Matrix that includes multiple stakeholders over the life cycle of a property, or a subset of it such as the Eco-Balance Sheet ™ that summarizes incremental costs and benefits of various projects so that we can weigh the value-added of different sustainable development projects in a whole new way. The following is a simplified version of the ECO-Balance Sheet ™ that categorizes both market and external costs and benefits of both use value and nonuse value; presented in a single table.

Total cost pricing—cont'd

ECO-BALANCE SHEET ™						
COSTS			**BENEFITS**			
MARKET	**EXTERNAL**		**MARKET**	**EXTERNAL**		
Financial	Environmental	Social	Financial	Environmental	Social	
Direct						<<< USE VALUE
Indirect						
OPPORTUNITY COSTS Foregone investment in another project or aspect of the project.			**OPPORTUNITY BENEFITS** Foregone consumption of energy and natural resources; reduced demand for building additional power plants; decreased allergic reactions and health risks.			<<< NON-USE VALUE
= TOTAL COST PRICING						

© Copyright Judi Schweitzer, All rights reserved.

The Eco-Balance Sheet ™ and Total Cost Pricing Matrix are applications of TCP. These tools offer a summary of both the "market" and "external" costs and benefits of each practice, component, or product under consideration to each of the many stakeholders in a given project. It facilitates the comparison of one project against another. "Market" costs and benefits are reasonably easy to identify and quantify. This example shows the initial investment in a photovoltaic (PV) installation (cost) and the estimated utility cost savings that might generate over the life of a home mortgage (benefit).

"External" costs and benefits are trickier to quantify at this point. Some represent common sense (water resources will be more scarce and costly in the future than they are now) but assigning a dollar value to that assumption is difficult. The greatest value of the TCP approach may be that it forces us to consider, even without all the numbers, the likely total positive and negative impact each decision might make on the project, community, and planet, so that we can decide which sustainable solutions produce the greatest value for the investment we are prepared to make.

Green Building is an essential component of sustainable development. It uses high performance technologies to reduce energy use, create healthier living environments, and preserve nonrenewable resources, and it typically refers to vertical construction.

A few developers have embraced and incorporated them with varying degrees of success. However, many remain skeptical, believing that these nonconventional approaches are more expensive than any pay-back they might generate. Despite a growing body of experience, evidence, and research, this resistance remains strong. Why?

Building truly sustainable projects and communities is not about piling on various green elements to build a score card or laundry list of accumulated green products and practices. We are learning that this approach can be prohibitively, and wastefully, expensive. We are also learning that understanding, respecting, and incorporating natural systems into our community master planning not only helps the resulting built environment perform better and more cost-efficiently but results in a community that is a more attractive place to live and work so that the developer creates more value and enjoys higher profitability. If we ignore the old adage "don't mess with Mother Nature," we do so at our peril.

More importantly, we are discovering that working in concert with nature and designing communities as living systems is actually not only less expensive initially than the traditional alternative but it creates greater value in the short and long term. This flies in the face of the widely accepted and persistent belief that sustainable development and green building practices are significantly more costly than conventional development.

A "whole systems" approach to cost/benefit analysis and value creation

Inherent in a concept of "value" is always a balance between investment and return. To arrive at a true concept of value, one must consider investment/cost as well as return/benefit, but a narrow interpretation of either side of the equation will throw off the balance. Until recently, the only development models at our disposal have relied on very narrow definitions of cost and benefit and their linear approach and assumptions led us to an inadequate and partial understanding of either side of the equation. By contrast, a whole systems approach enables us to evaluate components in a contextual way, to explore how they interact and work together. A whole systems approach leads us to complete understanding of both costs and benefits and ultimately to a better grasp of balance and [added] value.

All of us make value-based decisions every day. We decide to invest in higher quality roofing because it purportedly has a 30-year rather than a 20-year life. We buy an inexpensive electronic device made in a developing country because the technology will be outmoded soon and we will buy a different inexpensive device in a few years. If we had all the facts, though, and we knew the total costs to the planet of manufacturing that roofing material or electronic component (raw materials consumed, transportation costs, air and/or water polluted, hazardous waste created, etc.) we might have a different perception of "cost" and "value" and might make a different purchasing decision.

The concept of weighing total cost is not exactly new. As the World Resources Institute described in 1992 in The Going Rate: What It Really Costs to Drive, "The costs of driving can be categorized as either 'Market' or 'External.' Market Costs are those that are actually reflected in economic transactions .... Market costs represent the direct, the ordinary, expected costs of owning and operating a motor vehicle. In contrast, external costs (or 'externalities') are not reflected directly in market transactions ...., External costs obviously must be estimated using techniques other than analyzing normal market prices. Societal costs are the sum of market and external costs—in short, total costs."

As this quote makes clear, one pitfall we encounter when trying to understand "cost" in a sustainable context is that we are conditioned to equate "cost" with "price." We believe that the price we pay at the register, or the price at which we purchase a component on a per acre, per unit or per square foot basis (the "market" cost) represents its full cost. Because we assume that markets are effective and efficient, we also assume that the true cost is represented by the market price.

However, one needs only consider climate change and global warming to see a failed market (one that is out of balance), indicating that we have been living off natural capital versus natural interest, and to understand that "price" and "cost" are not the same. It is important to note that there are not only internal (market) and external (environmental and societal) costs and benefits; there are also both quantitative and qualitative aspects of value.

If we look at the other side of the equation, a narrow definition of "benefits" may also yield an inadequate perception of value. A simple calculation of ROI over a given period of time would not necessarily tell us the value of savings or higher sales prices that might be reaped, lower carry costs that might result, higher returns that could be attained, environmental benefits that might be realized, or negative impacts that could be avoided. It makes sense that if the cost of two components, for instance solar PV and low emissivity (low-e) windows, were the same, but one reduced air pollutants and greenhouse gases (GHGs) by a factor of 10 over the other, one might choose the component with the highest environmental benefit.

Interestingly, studies are showing that sustainably designed projects, from office towers to communities, lease or sell faster (and at higher per square foot prices) than comparable conventional projects in hot markets. Maximizing bottom-line returns requires intelligent use and allocation of financial and human resources and involves an excellent research and of series of decisions by numerous groups and stakeholders regarding a piece of real estate. Making informed decisions at each stage of the project is critical to increase the probability of success and maximize financial return of the investment.

Sustainable system versus linear approach

With a closed-loop sustainable systems approach (or whole systems thinking), we take effective decision-making about to a whole new level by better understanding the implications of our investment decisions. No one would intentionally create waste in a business; rather the goal is to run an efficient business with an effective use resources efficiently, eliminating waste in every area of the organization and construction process so that profits are not wasted, but maximized. The first step in accomplishing this is to better understand the difference between a linear system and a closed-loop sustainable system.

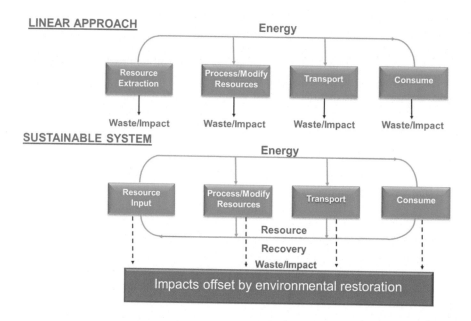

No savvy business person, who wanted their company to grow and prosper, would deplete the financial capital of a business for short-term return. Nor would any savvy business person allow "waste" throughout their business. Or do they? Given the recent scientific consensus about the data about global warming and climate change, which can be viewed as symptoms, it appears that we have been living off the earths "capital" rather than "interest" for our short-term gain, creating waste throughout the supply chain and life cycle of their projects. What do we do about it and how can we correct this path?

When we better account for all the elements of "cost" and all the elements of "benefit," we will reach a much fuller, and more sustainable, definition of "value." Fortunately, tools are emerging that will allow this more comprehensive approach to determining value added along with associated cost and benefits to find the appropriate value balance both for individual components and for an entire complex project. Instead of a linear analysis, they look more like loops, webs, or networks and use game theory and other techniques to analyze the effect that changing one variable in a system may have on many others.

Residential community developments are just beginning to be scrutinized systematically, in part because they have been slower to adopt green building practices than commercial and educational sectors and there are fewer examples to study. Studies of green office and school buildings by Greg Kats demonstrate that "Green Buildings provide financial benefits that conventional buildings do not." A report that analyzed 30 green schools built in 10 states during the period of 2001−2006 showed that an expenditure of $3-5 per square foot in green schools resulted in about

$70 net benefits per square foot, more than 20 times more than the cost of going green. "Conventional schools usually have lower design and construction costs and higher operational costs, whereas green schools usually have higher design and construction costs and lower operational costs." (Kats, Gregory. *Greening America's Schools — Costs and Benefits*, October 2006).

An additional investment of $3-5 per square foot in green elements in office buildings, based on a detailed review of 60 LEED rated buildings, confirmed that not only are they 25%–30% more efficient than conventional buildings, they produced $50–65 per square foot in net benefits over 20 years, over 10 times the additional cost of going green (Kats, Gregory. *Green Building Costs and Financial Benefits*, 2003). These studies demonstrate that savings resulting from energy, emissions, water, and even operations and maintenance, while undeniably important, are far from the most significant. Instead, financial benefits flowing from increased productivity, learning, and competitiveness coupled with reduced illness, asthma and allergies, absenteeism, turnover, and insurance costs are the most impressive and have the greatest financial benefits.

Evidence increasingly shows that, with a thoughtful investment of some additional time and money up front, a committed owner or master developer and a team that is creative and open to a systems-oriented design approach, green development does not have to cost more. In fact, it can cost less, both in the reasonably short term and certainly in the long term. Going green can actually pay a premium, especially when we consider, comprehensively, the total costs and realizable benefits of developing sustainably. Thanks to emerging tools and evaluation techniques, which are beginning to be able to prove this.

This chapter will examine some of the reasons developers and builders believe that adopting green practices, systems, and products is not cost effective. It will then discuss various methods of evaluating the total cost and total benefits of individual green elements and combined green packages as well as tools that can help a developer decide which green components will combine to make the most cost-effective mix in a given project—creating the greatest value. It is vital to understand that when we talk about green opportunities the initial cost is only a part of the picture. To understand the true economic impact of any green investment, it is imperative that we consider all its benefits to the project, community, and/or environment.

Part of the total value of any green decision is what is not spent, used, or wasted and what detrimental effects to air, water, vegetation, soil, quiet, human interaction, geopolitical balance, and other factors are avoided. Not having to build additional power plants or rectify a problem in the future, sometimes at huge cost, has a value today. The more we learn to identify and quantify all these near and long-term economic and environmental impacts, the better will we be able to make informed, defensible, and ultimately cost-effective and value-based choices on how to go green and how green to go.

Sustainable development costs more: perception versus perspective

Although sustainable development and green building have been around for a very long time, it appears that we are at a tipping point both in business and the global environment. For the first time, there is consensus in the international scientific community that the climate changes seen throughout the world are the result of human activity, and are not a natural occurrence. Observed climate changes and drastic increases in GHGs are definitively the result of human activities, primarily from the burning of fossil fuels. These are symptoms that we have been living off the earth's natural principal versus its natural interest.

For the first time, government, business, and consumers are taking note of the scientific community's concurrence that human activity is degrading the environment at a rate that cannot be allowed to continue. Today, consumers who value a lifestyle of health and sustainability account for 23% of the US population, and this is just the beginning. Legislation and building codes are greening the way we do business and in the way we design and build our communities. California is leading the way with legislation like the California Solar Bill (SB1) that mandates homebuilders to offer solar power as standard option by 2011, and on January 1, 2020, installation of solar PV became mandatory for all new low-rise residential units in California.

Research is proving that there is a premium paid by the consumer for highly energy efficient homes that make solar a standard feature. Incentives are being made to builders of new production homes. Financial institutions and insurance companies are giving discounts and incentives for projects that embrace environmental leadership. Companies that are viewed as environmentally responsible may garner greater brand equity. Companies that offer consumers an opportunity to be a part of the larger solution bring added value to their products.

A 2001 study by (Tilden Consulting, 2001) found that most developers believed at that time that green building costs between 10% and 15% more than conventional construction. Recent studies on Terramor Village at Ladera Ranch, a Rancho Mission Viejo project in California (Schweitzer, 2006) showed that residential builders were resistant to incorporating green elements and practices into their projects because they believed (1) there is a significant cost associated with green and (2) homebuyers would not pay whatever cost differential exists between those green and conventional homes, leaving the builder with a substantial "green squeeze" to his bottom line. A "high green cost" perception remains the predominant view of those that have not yet built a green project. Even with a generally successful experience installing mandated green practices and products at Terramor Village, most builders said that they would not change the way they design and build and were not planning to "green" their business.

Some of the early attempts at sustainable development may have given these practices a black eye because they suffered from the inevitable learning curve when any new paradigm, approach, or technology is implemented. Recent experience of those who are developing sustainably shows that an initial, and not insignificant, investment of time and effort in learning how to think differently is one of the costs of going green. Furthermore, a number of ill-fated, highly visible green development attempts only contributed to the stigma. Some trailblazing developers and builders tried to go green but experienced budget-breaking costs because they relied on green building "checklists" or point systems without a valid cost/benefit methodology, and/or the "add-on" of green components to an already designed or planned project. Projects that incur high additional costs may not have implemented "green thinking" early enough in the planning process or failed to calculate and capture savings that could have helped offset any additional costs, or both.

Widening our "Lens" on value

No one would run a business without accounting for its capital outlays. Yet most companies overlook one major capital component — the value of the earth's ecosystem services. It is a staggering omission; recent calculations place the value of the earth's total ecosystem services — water storage, atmosphere regulation, climate control, and so on — at $33 trillion a year.
Lovins, Amory B., Hunter L., Hawkins, Paul, 2000. A Road Map for Natural Capitalism, Harvard Business Review on Business and the Environment.

1) Earth from space

Reproduced from https://www.flickr.com/photos/gsfc/6760135001

Continued

Widening our "Lens" on value—cont'd

2) City aerial

Reproduced from https://www.pexels.com/photo/aerial-architecture-building-business-1343222/.

3) Street level view

Reproduced from https://www.pikrepo.com/fpkpr/low-angle-photo-of-high-rise-buildings.

Widening our "Lens" on value—cont'd

4) Drawings of building in plan view

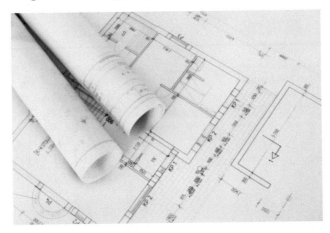

Is this the same building? In the first image, the building appears as a dot, in the third image, assuming the walls are parallel, they "appear to get closer together as do the mullions, and the floors appear to get closer together at a similar ratio–are the ceiling heights lower on the top floors? Are the walls really converting on the upper floors as well? In the fourth image, the building appears to be completely flat—is it?

Continued

Widening our "Lens" on value—cont'd

5) Science + Art = S+A

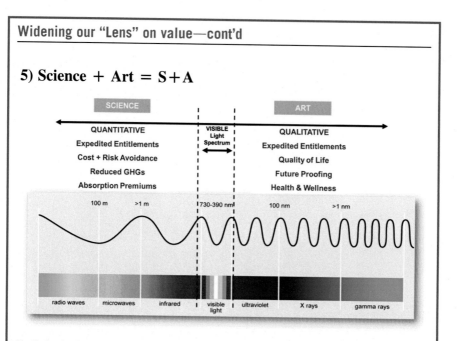

Similarly, the depiction of the light spectrum shows us that there are a great many more wavelengths of light than the human eye can see. Yet, these wavelengths are very much a part of our world and affect us all the time. The tenacity of the view that sustainable development unavoidably costs more than conventional development may be partially attributable to the way we "see," the perspective we have embraced as a result of our conventional training and experience.

If we were equipped with infrared night vision goggles, we could actually "see in the dark" because we could perceive the infrared rays in the spectrum that are not ordinarily visible. Much like infrared goggles, new tools enable us to see more of the total cost/benefit picture and hence take advantage of emerging market opportunities. A lot of the challenge of overcoming skepticism and resistance has to do with information and perspective.

Can you put a value on nature? There are a few ways to look at this (1) the value the natural ecosystems services in supplying heat, air, water, food, and shelter—our life support systems; (2) the value people pay for being close to nature and having open space incorporated into a master plan; and (3) the added value we as consumers place on products that are well designed.

Contrary to the old school wisdom that sustainable development costs considerably more than conventional development, the reality is that on average LEED Silver and Gold certified building cost only 1%–5% more over conventional construction—the other side of the equation is that of 10–20 times the value in cumulative utility savings, improved health, increased productivity, and employee retention. For production housing, recent green building projects have demonstrated only a 2%–5% higher cost range. Projects achieve these ranges by using methodologies that honestly assesses costs and savings, both initially and long term, as well as potential green revenues and premiums, resulting in a better understanding of value created. A number of projects have achieved very high levels of green building with no additional costs at all, or even cost reductions.

Terramor Village by Rancho Mission Viejo/DMB, Ladera Ranch, California

Recent studies on Terramor Village at Ladera Ranch (the largest solar community in the United States as of 2007), a Rancho Mission Viejo project in California, showed that residential builders were resistant to incorporating green elements and practices into their projects because they believed (1) there is a significant cost associated with green and (2) homebuyers would not pay the cost differential between those green and conventional homes, leaving the builder with a substantial "green squeeze" to their bottom line. (Schweitzer, Judi. *Making Green Pay,* Urban Land, May 2006).

Even with a generally successful experience installing mandated green practices and products at Terramor Village, most builders said that they would not change the way they design and build and were not planning to "green" their business. The homebuilders experienced total net green cost premiums averaging 5% net of utility rebates and state incentives (including direct and indirect costs including their learning curve). 75% of homeowners were willing to accept an increase of $25 per month and 35% were willing to accept more than $100 per month more in their mortgages if this equated to utility savings. Solar energy panels to generate electricity that reduce their monthly electric bill are essential or very important to 65%—91% depending on age. (Warrick, Brooke. Terramor Homebuyer Survey, AmericanLives).

Over a 30-year period, the solar PV panels installed on the 453 (67% of SFD) single family detached units are estimated to yield over $4+ million of energy generated onsite. Builders were still resistant insisting that once they sold the home, the cumulative benefits were lost to the homebuyer and they could not recapture the added first time costs. Direct construction costs were approximate. According to 7 of the 10 Terramor builders surveyed, their cost premiums (including directs, indirects, and additional costs due to their own learning curves) ranged from $2.02 to $11.40 (averaging $5.09 for SFD) and $6.61 for Mortgage Finance Authority (MFA). (Schweitzer, Judi. *Green Investing in Homebuilding: Lessons Learned at Terramor, Ladera Ranch, California*, Harvard GSD, AMDP, February 2006).

Overlaying the homebuilder and homebuyer data showed us that over 60% of the homebuyers were willing to pay more than it cost to produce. Overall interest for Terramor homes was 32% higher, and absorption 5% greater than homes in neighboring Ladera villages that lacked Terramor's green elements. Terramor homebuilders reaped a 5%—10% price premium over other homes in Ladera Ranch. Further study of the Village revealed that there were resale premiums for homeowners of homes with the solar PV reflected an even higher premium than was captured by the builders when compared to other comparable nongreen homes in Ladera Ranch. This begs the question—if builders widened their "lens" of "value," might they be able to see if there might have been missed opportunities by the Terramor homebuilders in communicating the value and cumulative benefits they built there, especially the homes with PV.

Carsten Crossings by the Grupe Company, Whitney Ranch, California

Carsten Crossings, a development of GrupeGreen at Whitney Ranch in Sacramento, offered highly energy efficient homes with PV as standard and they lucked out because the lots were coincidently oriented correctly for optimum sun angles.

Among the elements, the homes included as standard were solar electric and hot water, highly efficient air conditioning, double-glazed windows, and light tile (reflective) roofs with appropriate overhangs for shade.

The community started out to be a "Zero Energy Home" ((ZEH) a DOE Building America that is not "zero energy," but highly energy efficient, typically 50%–70% more efficient that a home built to code) and became a LEED for Homes (LEED-H) pilot project as well. The builder marketed these as highly energy efficient "solar homes." Clearly, the customer appreciated the value of these sustainable investments. In a down market, sales of the ZEHs outpaced all eight other housing projects within the Whitney Ranch master planned community by almost 2:1 (Whitney Ranch Sales & Traffic Report, April 23, 2006). In another example, at a Centex Homes project in Pleasanton, CA, the community where PV was standard sold at four times the velocity of a non-PV community.

Shea Homes, Scripps Highlands, San Diego, California

Not only do energy efficient and solar features increase sales velocity, they also support value over time or in down markets. A National Renewable Energy Lab study of Shea Scripps Highlands in San Diego, CA revealed that "energy-efficient homes and ZEHs not only hold their value, but increase their value at a faster rate than do conventional homes." ZEHs at Scripps Highlands enjoyed an average appreciation of 42% versus 26% for conventional homes in an adjacent community. (Comparative Analysis of Homebuyer Response to New Zero Energy Homes, NREL, 2004). Although it is too soon for many new green homes to reveal their long-term value appreciation, we have one example in Village Homes in Davis, CA, one of the first model solar communities, that was completed in 1982. As compared to adjacent communities in Davis, properties in Village Homes were selling at a premium of $11 per square foot in 1991.

As discussed earlier, the key to developing an economically viable sustainable community is to adopt a "whole systems" approach early in the planning process—a process that starts with natural systems, looks at the interconnections between building systems, and seeks solutions that address many problems within these systems. For example, the energy efficiency achieved through site orientation that works with the sun's angles and shell design that includes appropriate insulation and allows natural air flow can lead to the downsizing of an HVAC system and result in major net savings. A project that includes cogeneration or microgrids that reduce utility space needs between buildings may create more revenue-generating floor area at equal cost. Projects that pursue whole systems thinking are highly engineered, totally responsive to site characteristics and developed without following conventional building models. They represent the ultimate evolution of the sustainability-driven design processes.

If we attempt to patch green practices and products into a conventional development or building pro forma or program, chances are they *will* cost more. The way to achieve the most cost-effective green result is to start from the beginning with a different approach, perspective, and definition of value.

Examples of whole systems thinking with integrated solutions

As an example of whole systems thinking and integrated solutions, Pulte homes proposed a series of high performance energy conservation enhancements including an unvented roof, high performance windows, controlled ventilation system, and sealed combustion furnace that increased first cost by over $1600 per home in a project in Tucson, Arizona. By applying whole systems thinking, the developer was also able to produce savings of $1500 per home by installing a right-sized air conditioning system and avoiding installing roof vents, making the total net cost of the chosen green elements only $100 per home.

Another example is the Bentall Crestwood Corporate Centre in Richmond, British Columbia where the developer upgraded the building shell with low-E windows (low emissivity) windows, increased natural light and ventilation, and increased insulation R-factors. By using whole-systems thinking, the team was able to reduce the building's chiller from 200 to 50 tons, a cost reduction that more than offset the higher shell costs and resulted in a total building cost that was less than conventional construction.

To get the benefit of a whole systems approach, it is first necessary to assemble a multidisciplinary design and development team. It is also absolutely critical that the team initiate whole systems planning as early in the development planning as possible. The most successful sustainable projects are designed from the ground up (actually starting below the ground with the underlying hydrology), literally taking into account natural conditions (soil, drainage, wind, vegetation, solar orientation) and working with rather than against them. Generally, the later in the planning process any change is made, the higher its cost and the lower its potential to generate savings or efficiencies and increase profitability.

Implementing whole systems thinking requires a radically different approach to the design and development process. Those who have used it successfully note that identifying the interrelationships between components that lead to overall cost reductions requires considerably more time, intensity of effort, and special expertise on the front end. This may increase the investment in planning and design, making it seem disproportionately expensive, but these expenditures are more than offset by savings and benefits. The development team must be flexible and open-minded enough to allow certain line items to cost considerably more than in conventional developments with which they are familiar.

Life cycle assessment tools

The true value of sustainable design is realized over time. Making informed decisions for effective investments that optimize long-term sustainable value in community development requires appropriate "tools." Life cycle assessments (LCA), take into account all environmental costs and impacts associated with a product, including the raw materials, energy consumption, resource use and emissions through the entire manufacturing process, use and operation of the product during its projected life span, plus disposal or recycling costs. Although this "cradle to grave" level of inquiry is necessary to understand the full environmental costs and benefits of a product or practice, it is typically beyond most developers to implement in its purest sense. United States Green Building Council (USGBC) however is now attempting to integrate LCA into the LEED projects. In theory, LCA allows future cost savings to be traded off against higher initial capital costs; however, it is very difficult to quantify all impacts costs and benefits.

There is growing interest in LCA spurred by a number of efforts such as the U.S. Life Cycle Inventory Database project, the National Institute of Standards & Technology's BEES program, the Green Globes rating system, and the UNEP/SETAC Life Cycle Initiative also point to a growing interest in LCA (Building Design & Construction — Life Cycle Assessment and Sustainability, November 2005). Given the explosion of interest in mapping the total costs and impacts of extracting, manufacturing and using all the components of everything we take for granted in our daily lives, we can expect useful tools to emerge that will help us make many purchasing and use decisions in the future.

Eco-efficiencies and eco-effectiveness

Eco-efficiencies are so-called *solution multipliers,* or multiple savings, benefits, and/or reduced risks that can be realized "downstream" from an initial investment in a high performance component. Specifying a smaller chiller, for example, makes money available to upgrade the building envelope, while the improved building envelope allows that smaller chiller. These are first-cost savings due to anticipated downstream reductions in development and building costs. Too often, failing to identify secondary and tertiary cost savings offered by integrated sustainable products and practices leads to lost opportunities.

Although it may be tempting to select those green components with the lowest first costs, doing so does not optimize the value of long-term savings. A 2003 study by Greg Kats demonstrated that, on average, a 2% increase in upfront costs, if carefully invested, would, on average, result in a life cycle savings equal to 20% of total construction costs. ("*The Costs and Financial Benefits of Green Buildings: A Report to California's Sustainable Building Task Force*," October 2003).

Eco-effectiveness is a more sophisticated approach than "cradle to grave" realized through closing the loop of energy and waste with a "cradle to cradle" approach. Eco-effectiveness is further enhanced by linking interconnections of energy, resources, and waste. This concept was first presented by William McDonough, who underscores the important role of design in sustainability. The original concept was born out of the *Hanover Principles* written by McDonough and commissioned by the City of Hanover, Germany, conceived as a guide to the design of the 2000 World's Fair. "**Eliminate the concept of waste:** evaluate and optimize the full life cycle of products and processes to approach the state of natural systems, in which there is no waste." This concept was further expounded upon in his book, *Cradle to Cradle: remaking the way we make things,* (McDonough, William and Braungart, Michael. North Point Press, New York, NY, 2002).

Quantifying the green premium

Interestingly, studies are showing that sustainably designed projects, from office towers to communities, lease or sell faster (and at higher per square foot prices) than comparable conventional projects in hot markets. One key to capturing the full economic/financial benefit created by these green value enhancing elements is the ability to *quantify* and *explain* both "costs" and "benefits" for a truer picture of value. As an example, at Terramor Village where the master developer mandated certain green practices and products, we discovered that the home builders just assumed their customers would not value these green elements enough to pay a higher home price to get them.

The home builders invested little resources or creativity into either quantifying the potential energy, water or other savings homebuyers or home owners' associations might experience or educating the homebuyers on the value of these green features. A study of the priorities of homebuyers revealed that, even without education, over 55% of the actual homebuyers in that village would have been willing to pay more for their green homes in that green community than the aggregate cost of these items to the homebuilder. To realize the full benefit of any green practices or investments, it is necessary to educate a variety of audiences (community boards, financial sources, contractors and subcontractors, buyers) on their features and projected life-cycle performance.

It is also important to factor in the câché value attributed to sustainable development by end users. Most of us own a "designer" item or two for which we happily paid more than the intrinsic value because owning something branded had an appeal far greater than the sum of the cost of the raw materials and labor to produce it. Why would anyone buy a BMW, Mercedes Benz, or a designer suit if cost was the only metric? Particularly, in the residential sector, it is understood that consumers make decisions based on emotion as well as value. Sustainable communities that ground their residents in nature and are enjoyable, even inspiring, places to live and work are simply worth more to the families who flock there.

As Gerald Hines proved when he pioneered the use of signature architecture and superior building systems and components in high-rise office projects, tenants will pay more for a better environment. The value to the end user of good design, and increasingly of sustainable design, can be hard to quantify, but is undeniable. The emerging scientific discipline of the Neuroscience of Architecture is starting to better understand and document how our brain (and body) respond to the built environment. Some environments are proven to increase the body's ability to heal itself; proper lighting improves accuracy, etc. The General Services Administration is taking notice and is incorporating this growing body of knowledge in how they design and develop their properties.

Our experience and research have shown that one potentially significant benefit of green development—the green financing benefit—often missed and hence left on the table by developers or builders. As the financial, investment, and insurance industries seek to green themselves and their portfolios also seen as reducing risks to their investors, preferential mortgages, discounted property insurance, other boons are increasingly available to the savvy developer who is able to make a case for the green practices or components in a project. Moreover, municipalities are either increasing disincentives for conventional development and building like additional permitting restrictions, or adding new impact fees, or increasing incentives for green building and sustainable development, offering concessions such as streamlining of project permitting for green features, or both. A developer who borrows at a more favorable rate and then cuts his carry by decreasing his entitlement time and risk will see compounded benefits to his green investment. Discounted insurance rates make green building even more attractive.

Realizing the sustainability premium

The Sustainability Premium can be best understood when a developer effectively "accounts" for all direct and indirect costs and benefits of both "use value" and "nonuse" value for each of the stakeholders throughout the life cycle of the project and property. Capital costs of sustainable design can be similar or even lower than conventional rates due to lower operating costs that are starting to be recognized by lenders and insurance companies both in reduced qualifying levels as well as in reduced cost of funds.

Research is continuing to confirm that the market/consumers recognize the "value-added" of higher quality sustainably designed communities, homes, buildings, and products and respond with both their purchasing power and enhanced performance; consuming these products first when given the choice (in a down market), and in an up-market, they are willing to pay a "premium" compared to a "standard" product. Understanding all the benefits as well as all the costs and "accounting" for their "total value" is essential to making the maximum green our of sustainable community development.

In any sustainable community development, there are multiple stakeholders—master developer, builder, homebuyer, investor as well as those who derive economic benefit from all the sustainable practices and products involved in creating the project. When a project is designed and planned in an integrated fashion and when sustainable practices and components are incorporated all along the way, all the stakeholders share in the premium that is created. As this chart shows, the developer's community has more open space and is therefore more desirable and valuable. The homebuilder's product sells more quickly and at a higher price. The homebuyer has a more cost-efficient home that commands a higher resale price. A sustainable community, and the homes within it, are better than just green. The whole works together, in balance, and the value only grows over time.

Sustainability Premium is something that can be realized by multiple stakeholders to varying degrees depending on where the project is in an industry's life cycle. Research is confirming that the value-added premium is greater than the average profits of a standard project and equates to the cumulative quantitative and qualitative incremental benefits minus the incremental total costs. This can best be illustrated by plotting the property's life cycle (or time) along the X-axis and property's value along the Y-axis. The following is a conceptual representation (not to scale) of the expected Sustainable Value-Added Premiums identified for Rancho Mission Viejo's Ranch Plan based on research and analysis of Ladera Ranch and other neighboring master planned communities in Orange County, California.

Putting it all together

For most companies today, the only truly sustainable advantage comes from out-innovating the competition.

Moore, James F. May–June 1993. Predators and prey: a new ecology of competition.
Harvard Business Review.

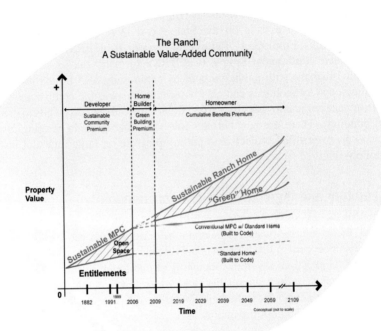

Financial wave in sustainability

Unsustainable buildings and communities will not be an asset in the future but will be a liability. The investment community has recently and increasingly recognizing the value of sustainable design and development. This includes pension funds, banks, mutual funds, corporations, and insurance companies to name a few. Capital is quickly lining up, at an increasing rate, to get behind development that is responding to environmental and sustainability concerns. Minimizing future risks and maximizing value is of primary concern to shareholders.

The Dow Jones Sustainability Indexes (DJSI) were launched in 1999 and track performance of top sustainable leaders globally. There are currently 60 DJSI licenses held by asset managers in 14 countries to manage a variety of financial products including active and passive funds, certificates and segregated accounts. In total, these licensees presently manage over 5 billion USD based on the DJSI. www.sustainability-indexes.com.

Other early efforts were spearheaded California in 2004 when the California State Treasurer Phil Angelides launched a landmark environmental "Green Wave" initiative to bolster improve long-term financial returns for pensioners and tax payers. "The four-pronged initiative calls on the State's two large public pension funds—the California Public Employees' Retirement System (CalPERS) and the California State Teachers' Retirement System (CalSTRS)—to marry the jet stream of finance and capital markets with public purpose by committing $1.5 billion to investments in cutting-edge technologies and environmentally responsible companies."

Insurance companies are also getting in line and offering discounts to green certified buildings. There are new entrants into the banking arena offering discounted rates to green certified projects and putting pressure on large financial institutions to follow suit.

Regulatory and legislative wave in sustainability

Going for the green, both in sustainability and profitability terms, may be its own reward. In the future, however, it may not be optional. In a very few years, projects that do not meet certain levels of sustainability may be considered white elephants or dinosaurs. Legislation governing standards of energy and water efficiency, carbon offsets/neutrality, indoor air quality, and waste management is proliferating and will eventually spread from hotbeds in California, the Northwest, Chicago, and the East Coast throughout the nation.

At the moment, meeting or exceeding standard building codes with sustainable practices may give projects a competitive edge in the market. Before too long, this paradigm shift into sustainable practice will become a requirement to being in the game at all. Integrated design and whole systems approaches will enable some projects to attain a level of excellence that is evidence of the highest creativity and entrepreneurialism, both hallmarks of the finest real estate development.

We are in the early stages of a paradigm shift that has been gaining momentum over the last 20 years. There seems to be a basic law of nature that at some point the cost of higher performance becomes disproportional to the added environmental benefit. In selecting which sustainability practices or products are incorporated into a sustainability program or project, it is first helpful to thoroughly understand the projects contextual framing (of the property/project/corporation) to effectively prioritize and consider goals, objectives, and measures within that context and overall sustainability continuum.

As previously described in Urban Land Institute's Developing Sustainable Planned Communities book, Steven Kellenberg and I explained that any master planned development has enormous numbers of variables. For the developer considering going green, or moving to a deeper shade of green, it may seem as if the variables multiply exponentially. Framing the project or company's sustainability or green initiative must consider context—maturity of the market and product, project scale, and bioregional climate are a few of the essential variables to understand. As each choice is made on how green to go throughout the development process, the developer is seeking the highest value and the greatest profitability for the shade of green she wishes to achieve. The optimum shade of green should evolve over time as the only thing you can count on in life and in business is change. The program needs to be anticipated changes to be sustainable. It is a rare project that can do everything to the highest level of sustainability.

To control costs and maximize potential financial gains from going green, most experts agree that the following are important:

- Early and decisive senior management leadership and commitment
- Competency and experience of planning and design team
- Clarity of goals and objectives
- Understanding local and regional [political, climactic, and market] conditions—conduct thorough research
- Application of both quantitative and qualitative valuation methodologies
- Optimize long-term value using appropriate life cycle costing tools
- Communication with and education of stakeholders including builders and contractors
- Flexibility in programming/criteria
- Providing sufficient time for planning and design
- Timing of bringing in builders/contractors
- Continued research and improvement of best practices

When contemplating a complex, multifacetted development project, it is useful to determine the spot on the continuum that best represents your goals and comfort level and use that as a filter for making each decision after first understanding the opportunities and constraints of the project context, scale, and timing. As you can see there is a band on the continuum that provides the opportunity for attaining the best balance between short term and long term and total costs and benefits creating the best balance between use and nonuse value of [financial, human, and natural] energy, and resources.

The push pull of regulation versus innovation

The single largest true cost savings "secret" seems to be the *timing* of integrating sustainable goals and design practices into the design process. The earlier in the planning such practices are introduced and evaluated, the higher the return on both soft and hard cost investments. If executed properly, the optimal market and project intervention is in the *Early Adopter Phase* of the *Technology/Innovation Adoption Curve* yielding the greatest Total Return on Investment with lowest life-cycle costs. In other applications, it has been described this as the "S-Curve of Innovation.[1]"

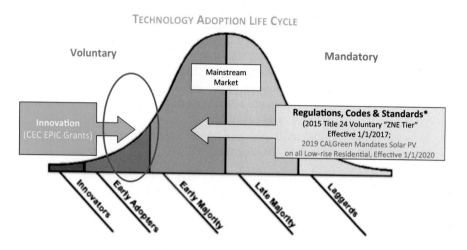

TECHNOLOGY ADOPTION LIFE CYCLE

Voluntary   Mandatory

Mainstream Market

Innovation (CEC EPIC Grants)

Regulations, Codes & Standards*
(2015 Title 24 Voluntary "ZNE Tier"
Effective 1/1/2017;
2019 CALGreen Mandates Solar PV
on all Low-rise Residential, Effective 1/1/2020

Innovators   Early Adopters   Early Majority   Late Majority   Laggards

*CA Long Term Energy Efficiency Strategic Plan Goal: All newly constructed homes to be ZNE by 1/1/2020; All new nonresidential buildings to be ZNE by 2030

References

Building Design & Construction, November 2005. Life Cycle Assessment and Sustainability.

Foster, R., 1986. Innovation: The Attacker's Advantage: Why Leading Companies Abruptly Lose Their Markets to New Competitors. And How a Few Have Avoided This Fate by Relentlessly Abandoning the Skills and Products that Have Brought Them Success. McKinsey & Co., Inc.

Katz, G.H., 2003. Green Building Costs and Financial Benefits.

Kellenberg, S., Schweitzer, J., 2007. Costs and Benefits of Sustainable Development, Developing Sustainable Planned Communities (Chapter 4). Urban Land Institute.

Lovins, A.B., Hunter, L., Hawkins, P., 2000. A Road Map for Natural Capitalism. Harvard Business Review on Business and the Environment.

MacKenzie, J.J., Dower, R.C., Chen, D.D.T., 1992. The Going Rate: What it Really Costs to Drive. World Resources Institute.

---

[1] Foster, Richard, 1986. Innovation: The Attacker's Advantage: Why leading companies abruptly lose their markets to new competitors. And How a Few Have Avoided This Fate by Relentlessly Abandoning the Skills and Products That Have Brought Them Success. McKinsey & Co., Inc.

McDonough, W., Braungart, M., 2002. Cradle to Cradle: Remaking the Way We Make Things. North Point Press, New York, NY.

Moore, J.F., 1993. "Predators and Prey: A New Ecology of Competition." Harvard Business Review 71 3, 75–86. Accessed December 6, 2009. Business Source Complete.

Schweitzer, J.G., 1993a. Eco-balance Sheet™. In: American Institute of Architects, World Congress of Architects, Sustainable Cities Conference Submission, Chicago, IL.

Schweitzer, J.G., 1993b. Closing the Loop with Sustainable Systems.

Schweitzer, J.G., February 2006. Green investing in homebuilding: lessons learned at Terramor village, Ladera Ranch, California. In: Harvard Graduate School of Design, Advanced Management Development Program - Independent Project.

Schweitzer, J.G., 2007. Property Values over Time: The Ranch — A Sustainable Master Planned Community.

Schweitzer, J.G., 2008. Applied Sustainability: Optimizing Value from Concept to the Boardroom to the Bottom-Line. Truman State University.

Tilden Consulting, 2001. Business Case for Green Building.

Value Based Management.Net. Rogers Adoption/Innovation Curve.

Warrick, B., 2006. Terramor Homebuyer Survey. AmericanLives.

Systems integrated mass transit to walking paths

3

Andrew DeWit, PhD

Professor, Kikkyo University, Toshima City, Tokyo, Japan

Consider Japan a leader in environmental and sustainability:
Sekisui House Leads Global Environment Award 2018.

Sankei Shimbun April 13, 2018

Sustainable Mega City Communities. https://doi.org/10.1016/B978-0-12-818793-7.00003-2

Sekisui House, Ltd—one of Japan's major high-quality home builders—was conferred the Grand Prize during the 27th Annual Global Environment Awards on April 9 for its eco-friendly, antidisaster "Smart-Community" Projects.

Four other companies were recognized for their environmental initiatives during the awarding ceremonies at the Meiji Memorial Hall in Moto Akasaka, Tokyo, in the presence of Prince and Princess Akishino. The awardees were given a certificate and trophy each.

The annual awards, sponsored by Fujisankei Communications Group (FCG), is given with the goal of encouraging industrial development compatible with the planet's environmental protection.

The Global Environment Awards and FCG-sponsored Grand Prize have been presented annually with the goal of recognizing businesses and other organizations for their innovations and activities conducive to curbing global warming and environmental conservation.

Sekisui House won the latest FCG Grand Prize for a range of work, including its role in constructing energy self-sufficient, disaster resilient "smart communities" in 16 locations across the country. It was also recognized for contributing to town-building design as shown by a project in collaboration with the city government of Higashi-Matsushima in Miyagi prefecture.

In the commendation ceremony, Sekisui House chairman Toshinori Abe said his company was determined to "continue to move ahead, throwing our corporate support behind the nation's regional environmental conservation programs for the sake of contributing to building a sustainable society."

The Sekisui-promoted initiative for the creation of what is known as Higashi-Matsushima City Smart Disaster Prevention Eco Town is Japan's first model for the local production and local consumption of renewable energy. The project provides electricity produced by solar energy to the whole city by making use effectively of smart grids—a next-generation transmission network for solar photovoltaic power generation.

The system in the city of Higashi-Matsushima, which is on the northeast coast of Japan's island of Honshu, has been highly evaluated. It is expected to reduce carbon dioxide emissions as well as show disaster resiliency by continuing to supply electricity for at least 3 days in the event of an emergency.

The FCG-backed Economy, Trade and Industry Minister Award was won by Kyushu Electric Power Co., Inc. The company, in cooperation with its utility groups, has been aggressively addressing the development and introduction of renewable energy sources such as geothermal heat and hydraulic power.

The Environment Minister Award was awarded to Toyo Ink SC Holdings Co., Ltd, for launching a lineup of biomass products ahead of any other firm, including an array of ink products for offset printing. The Land, Infrastructure, Transport and Tourism Minister Award was given to the Railway Systems Business Unit of Hitachi, Ltd., for the development of a next-generation rail car using a cutting-edge type of aluminum alloy material.

The Agriculture, Forestry and Fisheries Minister Award was bestowed on Sapporo Holdings Ltd, which pioneered the technology to produce bioethanol, an eco-friendly energy source, out of residues from the production of cassava tapioca starch. FCG chairman Hisashi Hieda noted in his opening remarks at the commendation ceremony, "Japan has been called upon to display strong leadership in tackling climate change since preparations went into full swing for formulating rules based on a new framework, effective in and after 2020, to implement the Paris Agreement, which stipulates specific measures for combating global warming."

Hieda also remarked, "We would like to contribute to striking a balance between the world's economic growth and countermeasures against climate change through these commendations." Fujio Mitarai, CEO and chairman of Canon Inc., and chairman of the screening committee of candidates for the Grand Prize for the Global Environment Awards, said at the ceremony, "We are firmly committed to broadening society's appreciation for these commendations in the cause for helping sustainable development of both society and economy."

Prince Akishino offered words of encouragement at the ceremony. Below are his remarks.

This Grand Prize for the Global Environment Awards marks the 27th year since its inauguration for the purpose of contributing to the benefit of society by commending activities achieving environmental conservation along with industrial growth. It is exemplified by such endeavors as the development of products and technologies that help materialize a sustainable, recycle-oriented society.

This commendation system has recognized a broad range of global environmental preservation tasks and heightened people's awareness of the need for environmental conservation. It has made significant contributions to addressing its goals by broadening the categories of those who are eligible for becoming award recipients—starting with the industrial world and then widening the scope to include municipalities, schools and civic groups. It honors those individuals and organizations who have been ardently engaged in environmental conservation activities.

Attention has been increasingly focused on a new international framework for the prevention of further global warming under the Paris Agreement and the Sustainable Development Goals adopted at the 2015 United Nations-sponsored summit meeting. In view of the circumstances, this nation, for its part, will likely be called upon to contribute further to the world through its superb environment-related technologies and knowledge.

The 27th Global Environment Awards Recipients: Grand Prize: Sekisui House Ltd.

METI Award: Kyushu Electric Power Co., Inc.
Environment Minister Award: Toyo Ink SC Holdings Co., Ltd.

MEXT Minister Award: Nogoya University.
Land/Transport Minister Award: Railway Systems Business Unit of Hitachi, Ltd.
Agriculture Minister Award: Sapporo Holdings Ltd.
Keidanren Chairman Award: Nippon Light Metal Holdings Co., Ltd.
Fujisankei Group Prize: YKK Corporation.
Honorable Prize: Toshiba Lighting & Technology Corporation.
_____Izumo Nishi High School Interact Club.

Better security and protection for people and ecological systems: integrated approaches for decoupling urban growth from emission pressures in megacities

Şiir Kılkış, PhD

Senior Researcher, The Scientific and Technological Research Council of Turkey, Atatürk Bulvarı, Kavaklıdere, Ankara, Turkey

Overview

The urgency of climate change clearly indicates that human civilization cannot continue as before, and entire systems, particularly urban systems, need to be open to realizing transitions that alter the ways of using energy, water, and natural resources. Urban emissions can be summarized based on three drivers, namely the use of embodied and operational energy, impacts of spatial decisions on human activities, and land use change (Seto et al., 2014). Urban systems require a focus on rapidly limiting and reversing these drivers of emissions while preserving precious carbon sinks toward reaching net-zero emissions. In particular, transitions in urban and infrastructure systems are necessary to accelerate the climate mitigation response for realigning existing trajectories to no more than 1.5°C of global warming (de Coninck et al., 2018). The scope of these transitions must take into account multiple urban systems in integration with urban planning. This chapter undertakes opportunities for decoupling urban growth and emissions from environmental pressures. The central focus includes perspectives from an original composite indicator while spanning urban growth and renewable energy solutions. Urban systems that achieve integrated approaches can secure a better future and contribute to a safer planet.

Need for reversing drivers of urban emissions in megacities

Changes in nighttime lighting that is visible from space are sufficient to underline the increasing trend of urbanization across time (Taubenböck et al., 2012). As

Sustainable Mega City Communities. https://doi.org/10.1016/B978-0-12-818793-7.00004-4

concentrated centers of human activity, cities have proportional demands on energy, water, and materials. Currently, cities are responsible for about 365 EJ of primary energy spending that corresponds to about 64% of the total spending worldwide (IEA, 2016), excluding the energy spending that takes place outside of urban areas to satisfy urban water and material demands, which in turn requires energy inputs. In addition, water supply for the 50 largest cities in the world results in a total distance of about 27,000 km (McDonald et al., 2014) that would be equivalent to more than half the equatorial distance of the world when placed in a straight line. Such vast distances have accompanied rapidly declining shares of the water supply from local sources, particularly in Beijing (Wang et al., 2019).

Strictly adhering to the common definition of a megacity, there are 33 megacities in the world as of 2018 with a population equal to or more than 10 million inhabitants (United Nations, 2018). Collectively, these megacities represent about 13% of the total urban population and 7% of the total urban and rural population. Considering land area including metropolitan areas as reported (Demographia, 2019), megacities represent less than 0.1% of the total terrestrial surface area of the Earth. This sheer density of human activity alludes to the demands of these urban areas on energy, water, and materials.

The identification of a megacity can change when the population is taken for the city proper, urban agglomeration, or metropolitan area. However, the most dynamic region based on the change in the number of megacities between now and the year 2030 will be Asia where 7 more cities are expected to join the existing 20 megacities in the region for a total of 27 megacities (United Nations, 2018). This is followed by Africa where 2 more cities are expected to be added by 2030 for a total of 5 megacities and Europe, which will have 1 more megacity for a total of 3 megacities.

Those in Latin American and the Caribbean and North America are expected to be stable while urbanization will continue without megacities in Oceania. By 2030, the number of cities with populations greater than 10 million is expected to increase to 43 cities. Considering these future projections, decoupling population growth from environmental pressures becomes more essential.

The individual cities are shown in Fig. 4.1 which contains the expected population growth in the year 2030 compared to 2018. There are also at least nine cities that are closest to the border line of 10 million inhabitants, including Xi'an and Chicago. Tokyo is the most populated city on the planet while New Delhi is expected to take the lead in 2030. New Delhi is currently also the most polluted megacity in the world based on annual mean particulate matter concentrations less than 10 μm in diameter (PM$_{10}$) at alarming levels, including 229 μg/m^3 (WHO, 2017).

Air pollution based on vehicular exhaust, industrial activities, and energy generation is responsible for multiple health impacts (Guilia et al., 2018). The temporary shutdown of thermal power plants and traffic limits in peak air pollution days is not a solution, and the megacity will continue to have serious issues if the energy system in and around the city does not change to renewable energy. Similar to New Delhi, multiple megacities are real case examples of the tragedy of the commons (Hardin,

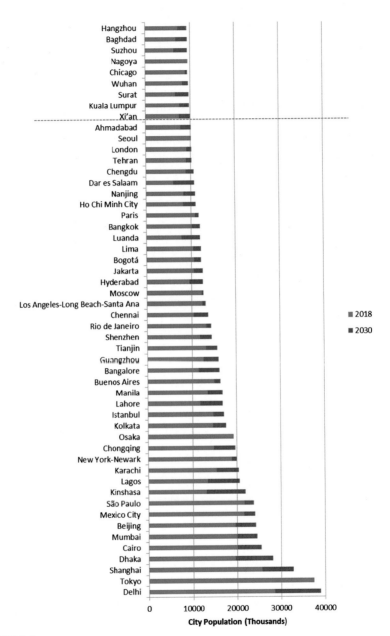

FIGURE 4.1

Comparison of cities with greatest number of inhabitants in 2018 and 2030.

Drawn based on data in The World's Cities in 2018 (United Nations, 2018).

1968). The most basic life-supporting systems, including air and water, are contaminated based on human activities due to natural resource usage.

The Atlas of Urban Expansion uses remotely monitored satellite data to discern certain urban metrics (NYU, 2019). In comparison to Fig. 4.1, 26 existing megacities have data on the average annual percentage change in the urban extent, which was greater than 2% for 80% of these megacities with an average of about 3.3% in the most recent 10–15 years. The cities with the highest annual average changes were Tianjin (13.7%) and Cairo (8.5%). The radar charts in Fig. 4.2 compare the

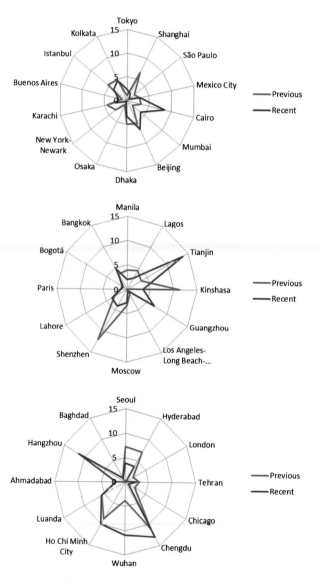

FIGURE 4.2

Comparison of the average annual percentage changes in urban extent.

Drawn based on data in the Atlas of Urban Expansion (NYU, 2019).

annual average percentage changes in the urban extent for the previous and most recent 10–15 years of remotely monitored data. Although the first two radar charts are for the 26 existing megacities with data in the atlas, the third radar chart represents those cities that are projected to become a megacity in 2030. Two data points from the previous 10–15 years are excluded for Guangzhou and Hangzhou that were above 20% while below 15% in the most recent timeframe. In contrast, scenarios for Tokyo based on compact and dispersed growth show implications on energy demand and renewable energy generation (Yamagata and Seya, 2013).

Multiple pressures from urban areas on natural resource usage need to be reversed in a time that is most critical to address global warming. Instead of urban population growth being followed by proportional increases in natural resource demands, including land use, cities must achieve better resource efficiency in a finite world. In any societal context, it is possible to view the concept of decoupling in absolute or relative terms. For absolute decoupling, environmental impacts should decline or at worst remain stable while desirable societal objectives are met. In relative decoupling, the impacts should be reduced per unit of output. Gross domestic product is a poor proxy for measuring societal well-being (Ward et al., 2016). Hence, another kind of decoupling between human welfare and additional demands on natural resources usage may be conceptualized in the process of improving the sustainable development of cities with co-benefits. Cities must purposefully and proactively decouple urban growth from environmental pressures.

It is possible to identify urban transformations as those leading to higher and lower levels of sustainability (Elmqvist et al., 2019). If this process was a race, then the winners could be cities that reach absolute decoupling in the shortest period of time while taking strides that differentiate their performance from any others on the track. Such cities would qualify as the most sustainable and harbingers of an ongoing urban transformation. In reality, megacities can be some of the most challenged cities among the typology of cities. Regardless of population, however, advantageous cities will be those that mobilize options for flexibility to realize paradigm shifts in urban energy, water, and environment systems with greater maneuverability in decision-making and the scale of implementation.

Benchmarked levels of sustainability performances of megacities

In total, 10 of the cities in Fig. 4.1 have been benchmarked with an original composite indicator to determine the relative level of sustainable development in the city with a particular focus on energy, water, and environment systems. The original composite indicator is the Sustainable Development of Energy, Water and Environment Systems (SDEWES) City Index with the namesake of the International Centre of SDEWES. The process of benchmarking is based on 7 dimensions and 35 main indicators as summarized in Fig. 4.3. There is also a subindicator framework that supports the main indicators as reported in previous publications (Kılkış, 2015, 2016, 2017, 2018a, 2018b, 2018c; SDEWES Centre, 2017). The SDEWES Index

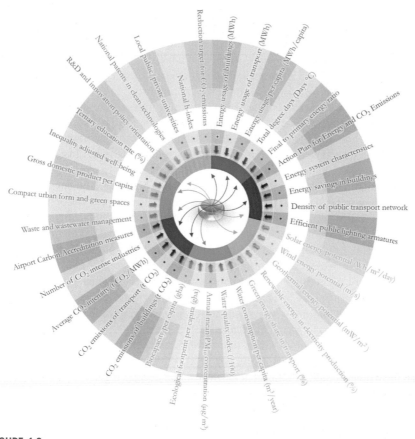

FIGURE 4.3

Seven dimensions and main indicator framework of the SDEWES Index.

Adapted from related publications and upcoming publication in JSDEWES

is applied to 120 cities (Kılkış, 2019a, 2019b), most of which are signatories to the Global Covenant of Mayors for Climate and Energy or have similar local strategies for climate mitigation.

In brief, the first dimension of the SDEWES Index on "Energy Usage and Climate" (D_1) requires a city to minimize energy usage in buildings and transport in the context of population, climate, and broader energy system efficiency. The second dimension on "Penetration of Energy and CO_2 Saving Measures" (D_2) requires a city to adopt measures for climate mitigation on the supply and demand sides, including an urban energy system that will enable the rational and efficient use of locally available energy sources, extending to renewable district heating and cooling networks whenever possible. The third dimension on "Renewable Energy Potential and Utilization" (D_3) necessitates that the city acts upon an awareness of the locally available renewable energy sources, including power and transport sectors. The fourth dimension on "Water Usage and Environmental Quality" (D_4) evaluates the current status of water usage and quality, air quality based on particulate matter concentrations as well as broader impacts on the environment based on demand for land that may lead to an ecological deficit or surplus.

In the fifth dimension on "CO_2 Emissions and Industrial Profile" (D_5), the CO_2 emissions of buildings and transport are the main concern along with the CO_2 intensity of the urban system as a whole and related factors. These include the presence of any CO_2 intense industries and/or renewable energy measures in airports servicing the city. The sixth dimension on "Urban Planning and Social Welfare" (D_6) compares the way that the urban system manages waste and wastewater and the spatial configuration to prioritize compact urban form and green spaces while upholding economic and educational opportunities.

The seventh dimension that covers "R&D, Innovation and Sustainability Policy" (D_7) complements all previous dimensions by assessing the status of any targeted orientation of R&D and innovation activities to address priorities in energy and environmental issues. Technological competence based on patents in clean technologies, including smart grids, and related aspects of the local ecosystem are further included in this dimension.

Fig. 4.4 maps the performance of 120 benchmarked cities against the percentage difference with the median value of an average city with 60 cities taking place above and 60 cities taking place below the central point (Kılkış, 2019a).

There are also four groupings, each of which contains 30 cities from the top performing cities that are characterized as the pioneering cities, which includes

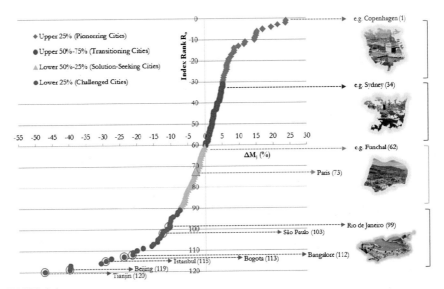

FIGURE 4.4

Performance of 120 cities that are benchmarked with the SDEWES Index.

Adapted from Kılkış (2019a) based on exemplary cities on the right and megacity markings.

Copenhagen, Stockholm, and Helsinki, followed by the transitioning cities. The transitioning cities include Sydney, which has recently adopted a renewable energy target. The Decentralized Energy Master Plan will be supported with urban–rural linkages, including bioenergy for anaerobic digestion (City of Sydney, 2013). Other groupings include the solution-seeking cities, such as Funchal that seeks to contribute to a smart island concept, and the challenged cities with multiple megacities.

There are multiple reasons for a given benchmarked city to take place in one of these groupings. Although the pioneering and transiting cities excel in multiple dimensions as the basis for an overall performance above an average city, this is not the case in other groupings of the benchmarked cities. The most drastic difference takes place in the last grouping of the challenged cities with below average performances in almost all dimensions for an overall performance that is well below those of an average city. From this analysis, the cities in the last grouping face multiple challenges and most notably, the urgent need to decouple growth from environmental pressures.

Energy usage, CO_2 emissions, and environmental quality, including air pollution, need to be addressed in a well-rounded manner if the city is to move toward greater sustainability in the future.

Among the 10 cities in Fig. 4.1 that are benchmarked with the SDEWES Index, 8 are current megacities and 2 cities take place among those that are projected to become megacities in the year 2030. As seen from Fi. 4.4, 7 of the 8 current megacities are benchmarked to have rankings in the last 30 cities that represent the challenged cities of the 120 city sample.

The two cities that are expected to become megacities in the year 2030, namely London and Nagoya, further take place among the challenged cities, including impacts due to the lack of a district energy infrastructure in London and the presence of industry in Nagoya. Other aspects include a relatively higher energy usage per capita due to urban affluence. There is only one current megacity, namely Paris, which takes place in the higher grouping of the solution-seeking cities with a rank of 73.

Table 4.1 summarizes the performance of the 8 current megacities in the benchmarked sample. Although Tianjin and Beijing have similar below average performances, one of the differentiating factors is the cogeneration-based district energy system characteristics of Beijing. However, due to the low level of penetration of renewable energy sources, it has not been possible to decouple energy usage from CO_2 emissions.

In fact, both Beijing and Tianjin perform more than 90% below the performance of an average city in the dimension that is related to CO_2 emissions (D_5). In contrast, the R&D and innovation competence of these cities are relatively high, which provides an advantage in changing the situation in the future. Already, advances in the development and deployment of renewable energy technology, as well as electric vehicles, are starting to become an advantage. Electric vehicle and bicycle sharing is also found to reduce gasoline usage and air pollutants (Sun et al., 2018) while greater shifts in sources of electricity are necessary.

Different from some of the other megacities, Istanbul does not perform significantly below the performance of the average city in the dimension that is related to renewable energy (D_3) mostly due to the share of hydropower in the electricity

Table 4.1 Performances across dimensions for megacities in the benchmarked sample.

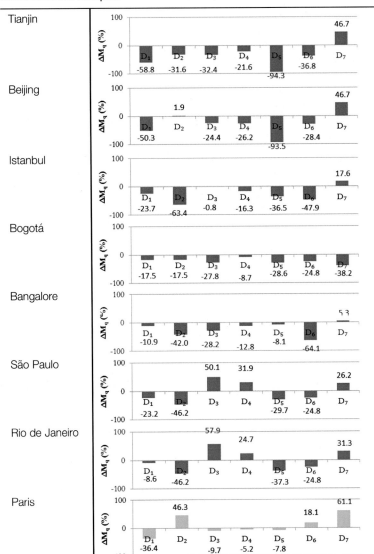

mix. However, the city performs significantly below others in the dimension related to energy and CO_2 saving measures (D_2) as the urban system lacks any district energy infrastructure for district heating and/or cooling. The building stock of Istanbul relies on the combustion of natural gas in heat-only boilers with fragmented use of cogeneration in certain commercial buildings and shopping centers.

Limitations of the sprawl of Istanbul in urban extent when compared to increases in population are also insufficient. Overall, this allows the megacity to have only one dimension in which the performance is above an average city, namely the last dimension (D_7) based on the presence of local institutes and R&D and innovation capacity that could be mobilized to address the needs of the urban system.

Extreme values that are below the performance of an average city are somewhat more limited in the performance of Bogotá as one of the cities from Latin America. At the same time, the city does not have a chance of performing above the average city in any given dimension also including the last dimension (D_7). This underlines even a greater role for cooperation with other cities to enable opportunities for improving the urban system.

In Bangalore, the most extreme value that remains below the performance of an average city is in the dimension that is related to urban planning and social welfare (D_6). Although the city has lower values of energy and water usage in the context of climate and population, such an advantage for certain indicators is not sufficient to overcome multiple other shortcomings, including aspects of air pollution, the recycling rate, wastewater management, and social welfare.

In contrast, the emergence of renewable energy solutions in several Indian cities, including photovoltaic panels in public transport stations, is starting to have examples in Bangalore, most prominently with the local airport installing solar energy (Krishnan and Dellesky, 2016). However, district energy infrastructure for district cooling is not considered. Greater penetration of renewable energy in Bangalore can provide multiple opportunities, including improved livelihoods.

As other challenged megacities, São Paulo and Rio de Janeiro have mixed performances across the dimensions of the index with those that are below the average city outweighing the others that are above. In aspects of the latter, the renewable energy utilization of both cities is relatively better than an average city mostly due to advantages in the hydropower share in the electricity mix and the use of cleaner fuels or electricity in transport, mostly based on bioethanol blends. However, the use of the renewable energy potential based on local energy generation in the urban context is not developed, which results in a below average performance for the penetration of energy and CO_2 saving measures (D_2).

Both Brazilian cities forego opportunities to benefit from possible efficiency improvements based on district cooling and opportunities to limit the urban extent to reduce the energy usage of transport based on urban planning. In addition, aspects of waste management are insufficient in both cities with recycling rates that are between 1% and 3%. In the case of Rio de Janeiro, the Visão 500 strategy that is prepared for the 500th anniversary of the city includes targets to increase the reuse and recycling of waste and increase ecoefficient buildings (City of Rio de Janeiro, 2017).

In the case of São Paulo, scenarios that represent the combination of energy and urban planning have indicated that 12% energy savings are possible while reducing emissions by 30% and increasing the renewable energy share by four times to reach 56% for all sectors (Collaço de Almeida et al., 2019a). The integration of aspects related to urban planning represented an improvement over another scenario for the megacity to increase energy self-sufficiency (Collaço de Almeida et al., 2019b). These opportunities are promising in raising the performance of these cities in the future. (Box 4.1).

Box 4.1 Additional narrative profiles of the challenged megacities

Tianjin is a port city that is about 30 minutes by high-speed train from Beijing. From the standpoint of public health, coal-fired power plants throughout China need to be limited, including those that are near to Tianjin (Zhang et al., 2019). The port is also the biggest port in the northern Yellow Sea Basin transporting half a billion tonnes of cargo each year, which supports the industrial base of the city. The measures that are implemented in the Sino-Singapore Tianjin Eco-City area with low energy buildings and a targeted recycling rate of at least 60% could have the potential of being scaled up and replicated in the city to address the need of decoupling growth from environmental pressures.

Notorious for sharing the characteristic of air pollution that is common in megacities, Beijing has taken dedicated measures to address this serious concern in recent years. The city banned coal burning and extended a cogeneration-based district heating network. The headquarters of the Chinese Academy of Building Research is one of the nearly net-zero buildings in the country. As a challenged city, however, energy spending and CO_2 emissions have not been decoupled that are aggravated by increases in urban growth. The public transport system is extensive based on a layered, concentric network while the magnitude of the commuting population still renders a severe traffic congestion problem.

Problems of water supply require extensive water transport distances. In addition, despite a striking growth in solar and wind energy in China, the power grid is still dominated by the use of coal power plants. The presence of universities and research institutes within the city can be a knowledge asset to move the city toward addressing some of the urban challenges. Existing studies include those on improving urban metabolism that could allow the city to better harness its environmental impacts.

Situated on two continents bridging Asia and Europe, Istanbul is also in the confluence of multiple urbanization challenges. The recent trend of building upward has significantly changed the urban landscape while higher rise buildings require greater embodied energy due to the use of stronger structural materials and any decorative façades. One study for urban development in Turkey indicates the need to minimize emissions based on multiple drivers, including embodied energy use in buildings (Kılkış and Kılkış, 2019).

The built environment of the city relies on the use of natural gas directly for heating demands in buildings, which represents a mismatch in the useful work potential of the energy resource based on the second law of thermodynamics (high quality, high exergy) and the low exergy demand for space heating. The city lacks an integrated approach to urban planning without ambitions for district energy networks that could supply energy more efficiently to the built environment with possibilities for the integration of renewable energy sources. Modes of cogeneration are restricted to individual sites, such as shopping centers, rather than the district scale. The recycling and composting rate of the city is 2% although a recent initiative for zero waste is launched at the national level.

Progress toward decoupling ongoing urban growth from increasing environmental pressure requires systemic changes in the ways that urban planning is integrated with the urban sectors. One study that is put forth for São Paulo asked the question of whether this megacity can become a megacity with greater use of endogenous energy resources (Collaço de Almeida et al., 2019b). The answer included contributions from solar energy utilization, municipal solid waste and manure, biomass from pruning, and micro-hydro. In addition, higher renewable energy shares with and without benefits from urban planning were undertaken (Collaço de Almeida et al., 2019a).

The megacity can reduce greenhouse gas emissions more when aspects of both renewable energy and urban planning are taken into account. Such scenarios are promising to enable greater energy and CO_2 emission savings in São Paulo and particularly for realizing a relative or absolute decoupling in megacities. Rio de Janeiro has been the home venue of the Earth Summit in which the need for "sound urban management" for all cities of the world was first adopted in the context of an international agenda, namely Agenda 21 (United Nations Division for Sustainable Development, 1992). Rio de Janeiro continues to be symbolic of the urban challenges in developing countries.

Continued

BOX 4.1 Additional narrative profiles of the challenged megacities—cont'd

From the aspects of the urban energy system, the cooling loads are met with split systems, which pose additional demands for electrical energy that could be allocated to electric mobility. Although the renewable energy share in the energy mix is higher than Asian or Eurasian megacities, one of the weaknesses of Rio de Janeiro is waste and wastewater management. One recent attempt to alter this situation was the launch of the Visão 500 local strategy document that included a revitalization of the bays of the city by increasing water quality, including the Bay of Guanabara and the lagoon of Barra. Exemplary opportunities to improve performances of Rio de Janeiro are discussed in Kılkış (2018b). More efficient utilization of resources can also open possibilities for greater well-being and enhanced livelihoods, including cooperatives in the recycling sector (Tirado-Soto and Zamberlan, 2013).

As another megacity in South America, Bogotá shares certain shortcomings in urban infrastructure while greater urban—rural linkages, including the possible use of banana and coffee pulp waste for bioenergy generation (ESMAP, 2014), could provide opportunities to increase the renewable energy share. This could also provide a means to valorize agricultural waste and provide additional revenue in the local economy while displacing the need for additional primary energy spending. The annual mean solar energy potential of both Bogotá and Bangalore between 4811 Wh/m^2/day and 5930 Wh/m^2/day also exceeds those of some other cities. Currently, Bangalore Kempegowda International Airport has a 3.4 MW solar installation for on-site generation. This is expected to increase to 12 MW by 2020 while sourcing another 8 MW from off-site to reach climate neutrality. Such progress could be implemented in other urban areas.

The only existing megacity that has a performance better than the challenged cities of the sample is Paris, which takes place among the solution-seeking cities as the next grouping of 30 cities in Fig. 4.4. One of the dimensions in which Paris has clear advantages includes the penetration of energy and CO_2 saving measures (D_2) based on a well-developed district energy infrastructure. The district heating network has a 53% renewable or recovered energy share based on 12% bioenergy, 1% geothermal, and 39% waste. In addition, the expansion of a 71 km district cooling network has been enabled by utilizing the existing sewage network (UNEP, 2017).

The affluence of the city limits reductions in energy usage per capita, however, which provides one disadvantage for the indicators on energy usage (D_1). The performance of three dimensions from D_3 to D_5 are near average, including due to a relatively low renewable energy share in the electricity mix (D_3) while geothermal energy is extensively used in Paris-Orly Airport to meet 70% of heating demands. The last two dimensions have above average performances, including those based on a negative sprawl index, which compares population and urban area, about a 23% share of green areas in the city, and an innovation ecosystem that is capable of supporting the climate targets of the city. The city targets a 15% renewable energy share in the city and energy savings of 35% by 2030 (de Paris, 2018).

In contrast to other megacities, Paris has taken greater strides to decouple urban growth from environmental pressure, including a district energy infrastructure that utilizes multiple sources of renewable and recycled energy, while additional progress is necessary. The import of any biomass pellets requires attention to reduce the ecological footprint. Most recently, the city has also ordered 800 electric vehicles for the public transport fleet. Higher penetration of renewable energy sources in the energy system is necessary to increase the benefit of such a switch. These include the use of these assets to increase the flexibility of the energy system with a greater share of intermittent renewable energy sources and benefiting from the greater employment benefits of renewable energy.

Beyond the benchmarking of cities with a focus on the sustainability of energy, water, and environment systems, the SDEWES Index is used to identify collaboration pairs based on cities with similar performance across the dimensions. In contrast to opportunities for policy learning across cities on an ad hoc basis (Marsden et al., 2011), the collaboration pairs can provide a more targeted approach to addressing joint challenges collectively.

Based on a process of pairing cities with above and below average performances in the same dimensions across the index (Kılkış, 2018a), four pairs of cities that involve at least one of the eight existing megacities are formed among a search across all 120 cities. These collaboration pairs with similar above and/or below average dimension performance as a representation of similar urban challenges can be partly followed from Table 4.1. Two of these pairs include only megacities, namely (1) Tianjin, Istanbul, and Bangalore and (2) São Paulo and Rio de Janeiro. The other pairs also involve cities that are not megacities, namely (3) Bogotá and Johannesburg and (4) Paris, Incheon, Nagoya, and Antwerp.

Currently, the top performers in the SDEWES Index are Copenhagen, Stockholm, and Helsinki (Kılkış, 2019a). These cities have urban systems that are advanced from multiple perspectives, including innovative concepts in district energy infrastructure. Already in the suburbs of Copenhagen, pilots toward a fourth generation district heating network are being undertaken with energy savings due to a lower-temperature network (Danish Energy Agency, 2017). In addition, through the concept of open district heating in Stockholm, the city is supporting its target of phasing out the use of fossil fuels in the district energy system based on contributions from sources of waste heat in the urban area, including data centers. By 2022, it is expected that 10% of the remaining share of fossil fuels will be substituted in this way (Stockholm Exergi, 2019).

In addition, the waste of the city is being used to produce biochar, biomethane for waste trucks and energy while reducing the amount of waste that is generated in the city (Stockholm Executive Office, 2016). Helsinki is another Scandinavian city with a high level of performance across the dimensions. In the future, studies on the sharing economy have also shown that the use of shared mobility can drastically reduce the need for private vehicles to about 4% while maintaining the same level of mobility (OECD, 2017).

The cities of Espoo, Århus, and Gothenburg follow the top three performing cities (Kılkış, 2019a). The advantages of Gothenburg include urban planning to increase compact urban form (Cereda, 2009) and limiting urban sprawl while increasing the share of permeable surfaces as a water-wise city (IWA, 2018). The district heating network is also used to replace the use of bunker fuels in ships that are docked in the port area (SCIS, 2019).

In contrast to megacities, one of the models for urban development may be to pursue more sustainable, smarter, and balanced growth with various cities of more limited sizes.

In one of the future visions for the region, a plan for a virtual city with about eight million inhabitants connecting Copenhagen, Gothenburg, and Oslo was put forth based on the idea of establishing a high-speed train connection between these cities of about 1 hour on either side of Gothenburg (Region Skåne, 2014). Additional connections to Stockholm and Hamburg further increased the concept to 12 million inhabitants. Eventually, the plan was not pursued.

However, such an idea requires reflection on what may shape the future of sustainable megacities. Concentrations of population greater than 10 million in a single location can be utilized to provide urban systems and infrastructure that may decouple urban growth from environmental pressures. At the same time, balancing the trends of urbanization across existing cities while minimizing impacts on land use may also be used to reconcile the concept of megacities and sustainable development in the future.

Our Planet even when viewed from space appears as a fragile home and is greatly becoming more so due to the risks of nearing tipping points in the global climate (Steffen et al., 2018). The best chance of increasing the sustainability of urban systems will include a significant role for tackling the urban challenges of existing megacities and megacity communities that constitute major demand points for energy, water, and material usage. Simultaneously, finding solutions that will allow other cities to avoid the same challenges as megacities will require significant effort. Clearly, solutions for sustainable urbanization big or small will receive prominent support based on renewable energy solutions across urban sectors, aspects of integrated spatial planning, and coherent policies to ensure the sustainable development of urban systems. The combination of these elements in an integrated approach will provide the best opportunities to tackle the drivers of urban emissions.

Achieving decoupling through renewable energy penetration

An important advantage for a city to have a high level of performance in the SDEWES Index is to adopt an integrated approach based on the penetration of renewable energy. When coupled with energy savings, this strategy of decoupling will provide an advantage across multiple indicators and dimensions. Energy savings should also be integrated with aspects of better urban planning, which can enable structural changes, such as zoning for district heating and/or cooling networks, and impose behavioral changes, such as avoided car journeys in walkable neighborhoods and transit-oriented development. At the same time, synergies across sectors are important, as the penetration of renewable energy sources in the energy system is most successful when sectoral linkages are formed.

Even the water sector can provide opportunities for demand response to avoid the curtailed use of electricity from intermittent renewable energy sources due to mismatches in the time of supply and demand (Duíc et al., 2008). In arid areas that require desalination to meet water demands, the storage of fresh water and brine can act as a medium for flexibility to increase the utilization of renewable energy (Novosel et al., 2014). In addition, urban water systems can integrate renewable energy in the water supply infrastructure and use wastewater for both biogas and waste heat (IWA, 2019).

Most recently, an energy scenario based on 100% renewable energy utilization at the global level has been modeled based on the integration of the power, heat, transport, and desalination sectors (Ram et al., 2019). Such a scenario is given to be consistent with the level of transformation that would be necessary for a 1.5°C scenario. As co-benefits, 35 million new jobs will be enabled by a shift to the 100% renewable energy scenario and energy cost savings across all regions of the world by 15%−34% (Ram et al., 2019).

Given the prominence of urban systems in the global energy system, cities and especially megacities have a major role in delivering upon these opportunities while integrating sectors and accelerating the global response to mitigate global climate change. Integrated, smart energy systems at the urban level can also enable 100% renewable energy solutions with viable demands on local sources (Bačeković; Østergaard, 2018).

In the United States, scenarios for 53 towns and cities based on 100% solar, wind, and wave energy have indicated employment and well-being co-benefits based on cleaner air and the environment (Jacobson et al., 2018). Decoupling urban impacts with renewable energy penetration is a solution for both present and future generations in a healthier planet.

As of January 2018, 42 cities of various sizes self-reported to obtain 100% of their electricity from renewable energy sources while 59 cities had at least a 70% share and 22 cities had at least a 50% share (CDP, 2019). Among the megacities, those that are included in this reporting with at least a 50% of the electricity coming from renewable energy sources are São Paulo (50%) and Dar es Salaam (70%). Although there are multiple cities from each inhabited continent, Asia currently has one city in the scope of the study, which is Inje (70%).

In contrast, as an outlook for the future, multiple cities from every continent adopted 100% renewable energy targets (Renewables 100 Policy Institute, 2019). Most notably from Asia, Fukushima has adopted a 100% renewable energy target by the year 2040. In Oceania, Canberra targets becoming a renewable powered capital by 2020 and Sydney has adopted a 100% renewable energy city target.

In Europe, Edinburgh adopted a 100% renewable energy target, which is notable in the context of a national net-zero target that passed a first vote (SSN, 2019) and a declaration of a climate emergency. Numerous other European cities have already reached at least a 50% share of renewable energy on their way toward a 100% renewable energy target. Reykjavík achieved this level in the power sector and Copenhagen aims to be the first climate neutral city in the world as an enabler of a 100% renewable energy target at the country level by 2050. The capital region with 33 municipalities has also adopted a target to decarbonize the power and heat sectors by 2035 and the transport sector by the year 2050 (Klima100, 2018).

The city of Klagenfurt within the region of Carinthia aims to be a 100% renewable energy region. In addition to Copenhagen, Reykjavík and Klagenfurt also take place among the top 10 cities that are benchmarked in the SDEWES Index among the pioneering cities. The complete ranking of the benchmarked cities can be found in a publication in *Renewable and Sustainable Energy Reviews* (Kılkış, 2019a).

The framework of the SDEWES Index is useful not only for benchmarking the present status of cities but also for putting forth an integrated perspective that can be used to plan strategies for decoupling urban growth from environmental pressures. First, the effective combination of energy efficiency and renewable energy across urban sectors, including district energy systems for power and heat/cold, low energy buildings, public and electric transport, and water sectors will enable cities to attain simultaneous improvements in multiple dimensions, including aspects of energy savings (D_1), mitigation measures (D_2), renewable energy utilization (D_3), improvements in air quality (D_4), and significant decreases in CO_2 emissions with possibilities to attain net-zero emission levels (D_5).

Moreover, it is important to couple an increase in the renewable energy utilization of the urban system based on aspects of urban planning (D_6). More compact urban form can enable greater opportunities for district energy infrastructure. Mutual shading of buildings may reduce solar energy utilization while this can be managed in various ways (Lobaccaro et al., 2019). The planning of mixed-used buildings can be used to balance the supply and demand of energy to diversify concurrent energy loads (Wu et al., 2018).

Sustainable development is about capturing synergies and co-benefits. For example, renewable energy solutions will increase livelihoods and economic opportunities in and beyond cities. Associated reductions in conventional forms of energy generation can even extend to improved water quality levels due to reductions in water stress. In aspects of an energy-water nexus, processes for conventional forms of energy generation are water intensive, including fossil fuel extraction, thermoelectric power generation, evaporative losses in hydropower dams, and bioenergy if not based on crop residues (Meng et al., 2019).

Opportunities for increasing circular economy in the urban context to valorize waste in different forms, including biowaste, will lower demands on natural resources and boost value generation. The R&D and innovation capacity of the city needs to be directed to supporting all of these available options to improve the sustainability of urban systems through the upscaling of pilot sites as represented in the scope of D_7.

Overall, obtaining improvements in the SDEWES Index requires a well-rounded approach across urban sectors, urban planning, and mobilization of the local ecosystem in alignment with the aims of urban sustainability. From a governance perspective, the regulatory power of local authorities is one of the available modes of urban climate governance (JRC, 2018). Local authorities govern through regulations in addition to enabling partnerships, providing incentives and resources, and implementing strategic investments in municipal assets.

Local and higher authorities have a greater role to ensure that the rules are aligned with the urgency of increasing urban transformations across multiple urban sectors and urban planning. Related regulations can include enabling energy savings and renewable energy penetration for decarbonizing the heating sector through mandatory connections to district energy networks based on zoning and bans on inefficient technology. Copenhagen will also completely eliminate diesel buses by 2025 and the United Kingdom will ban natural gas boilers in all new homes by the year 2015.

The vision of a SDEWES Aware City is defined as a city that "uses energy resources rationally at the right amount, quality and time toward 100% renewable energy systems, acts to preserve water resources and seeks integration whenever possible to valorize limited natural resources in respect of environmental balances" (Kılkış, 2019a).

Such a vision is highly relevant for the kind of bold and transformative action that is necessary to achieve any scenario that is more capable of moving the progress of human civilization closer to the much safer 1.5°C target for limiting global warming. The ability to realize the 1.5°C target with at least a 66% probability of not exceeding this value requires that CO_2 emissions are reduced by about 40%−50% by the year 2030 from 2010 levels (Rogelj et al., 2018) while time is running short for a strong urban renewable energy response.

The vision for a SDEWES Aware City requires that strategies for climate mitigation capture synergies between urban sectors to reach the greatest potential for limiting energy usage and CO_2 emissions in the limited time ahead. To reach a level of accelerated transitions across energy, water, and environment systems, a SDEWES Aware City must also implement an effective set of policy tools to enable a governance structure that can move away from a fragmented view. Additional investments that would continue to lock-in CO_2 emissions need to be eliminated based on broader urban systems thinking.

The bold and transformative action that is necessary for climate mitigation holds opportunities for realizing greater penetration of renewable energy across the urban system, greater integration including across urban planning and better coherence in governance structure. These aspects can guide progress in the right direction with the needed pace for realizing transitions across the urban system as a whole. Only through the realization of such urban transitions will it be possible to enable more sustainable megacities and megacity communities that can effectively decouple environmental pressures from urban growth. Urban action is crucial to secure a better future for people and the planet as a whole.

References

Bačeković, I., Østergaard, P., 2018. A smart energy system approach vs a non-integrated renewable energy system approach to designing a future energy system in Zagreb. Energy 155, 824−837.

CDP, 2019. The World's Renewable Energy Cities. https://www.cdp.net/en/cities/world-renewable-energy-cities.

Cereda, V., 2009. Compact City and Densification Strategies: The Case of Gothenburg (M.Sc. thesis). Blekinge Tekniska Högskola, Karlskrona, Sweden.

City of Rio de Janeiro, 2017. The Rio of Tomorrow: Vision Rio 500 and Strategic Planning 2017-2020 (in Portuguese), pp. 1–399.

City of Sydney, 2013. Decentralised Energy Master Plan Renewable Energy 2012–2030.

Collaço de Almeida, F., Simoes, S., Dias, L., Duic, N., Seixas, J., Bermann, C., 2019a. The dawn of urban energy planning — synergies between energy and urban planning for São Paulo (Brazil) megacity. Journal of Cleaner Production 215, 458–479.

Collaço de Almeida, F., Dias, L., Simoes, S., Pukšec, T., Seixas, J., Bermann, C., 2019b. What if São Paulo (Brazil) would like to become a renewable and endogenous energy -based megacity? Renewable Energy 138, 416–433.

Danish Energy Agency, 2017. Danish Board of District Heating - State of Green. District Heating - Danish Experiences.

de Coninck, H., Revi, A., Babiker, M., Bertoldi, P., Buckeridge, M., Cartwright, A., Dong, W., et al., 2018. Strengthening and implementing the global response in Press. In: Masson-Delmotte, V., Zhai, P., Pörtner, H.O., Roberts, D., Skea, J., Shukla, P.R., Pirani, A. (Eds.), Global Warming of 1.5°C, an IPCC Special Report. https://www.ipcc.ch/sr15.

de Paris M. 2108. City of Paris Passes its New Air Quality. Energy & Climate Action Plan.

Demographia, 2019. World Urban Areas 15th Annual Edition: 201904.

Duić, N., Krajačić, G., da Graça Carvalho, M., 2008. RenewIslands methodology for sustainable energy and resource planning for islands. Renewable and Sustainable Energy Reviews 12 (4), 1032–1062.

Elmqvist, T., Andersson, E., Frantzeskaki, N., McPhearson, T., Olsson, P., Gaffney, O., Takeuchi, K., Folke, C., 2019. Sustainability and resilience for transformation in the urban century. Nature Sustainability 2, 267–273.

ESMAP Energy Sector Management Assistance Program, 2014. Tool for Rapid Assessment of City Energy — Bogotá, Colombia.

Gulia, S., Mittal, A., Khare, M., 2018. Quantitative evaluation of source interventions for urban air quality improvement - a case study of Delhi city. Atmospheric Pollution Research 9 (3), 577–583.

Hardin, G., 1968. The tragedy of the commons. Science 162 (3859), 1243–1248.

IEA, 2016. Energy Technology Perspectives 2016 — towards Sustainable Urban Energy Systems. OECD/IEA, Paris.

IWA, 2018. City Water Stories: Gothenburg. http://www.iwa-network.org/city/gothenburg/.

IWA, 2019. The Roadmap to a Low-Carbon Urban Water Utility.

Jacobson, M.Z., Cameron, M.A., Hennessy, E.M., Petkov, I., Meyer, C.B., Gambhir, T.K., Maki, A.T., et al., 2018. 100% clean and renewable wind, water, and sunlight (WWS) all-sector energy roadmaps for 53 towns and cities in North America. Sustainable Cities and Society 42, 22–37.

JRC, 2018. Guidebook: How to Develop a Sustainable Energy and Climate Action Plan (SECAP), Part 3 Policies, Key Actions, Good Practices for Mitigation and Adaptation to Climate Change and Financing. https://ec.europa.eu/jrc/en/publication/eur-scientific-and-technical-research-reports/guidebook-how-develop-sustainable-energy-and-climate-action-plan-secap.

Klima100, 2018. 100 Climate Solutions from Danish Municipalities. Sustainia-Realdania.

Krishnan, D., Dellesky, C., 2016. Bangalore's Airport to Become a Leader in Solar Energy Production. WRI Ross Center for Sustainable Cities.

Kılkış, Ş., 2015. Composite index for benchmarking local energy systems of Mediterranean port cities. Energy 92 (3), 622—638.

Kılkış, Ş., 2016. Sustainable development of energy, water and environment systems index for Southeast European cities. Journal of Cleaner Production 130, 222—234.

Kılkış, Ş., 2017. Benchmarking the sustainability of urban energy, water and environment systems with the SDEWES city index and envisioning scenarios for the future. In: Plenary Lecture 12th SDEWES Conference Dubrovnik, October 4—8, 2017.

Kılkış, Ş., 2018a. Application of the sustainable development of energy, water and environment systems index to world cities with a normative scenario for Rio de Janeiro. Journal of Sustainable Development of Energy, Water and Environment Systems 6 (3), 559—608.

Kılkış, Ş., 2018b. Benchmarking South East European cities with the sustainable development of energy, water and environment systems index. Journal of Sustainable Development of Energy, Water and Environment Systems 6 (1), 162—209.

Kılkış, Ş., 2018c. Sustainable development of energy, water and environment systems (SDEWES) index for policy learning in cities. International Journal of Innovation and Sustainable Development 12 (1/2), 87—134.

Kılkış, Ş., 2019a. Benchmarking the sustainability of urban energy, water and environment systems and envisioning a cross-sectoral scenario for the future. Renewable and Sustainable Energy Reviews 103, 529—545.

Kılkış, Ş., 2019b. Data on cities that are benchmarked with the sustainable development of energy, water and environment systems index and related cross-sectoral scenario. Data in Brief 24, 103856.

Kılkış, Ş., Kılkış, B., 2019. An urbanization algorithm for districts with minimized emissions based on urban planning and embodied energy towards net-zero exergy targets. Energy 179, 392—406.

Lobaccaro, G., Croce, S., Lindkvist, C., Munari Probst, M., Scognamiglio, A., Dahlberg, J., Lundgren, M., Wall, M., 2019. A cross-country perspective on solar energy in urban planning: lessons learned from international case studies. Renewable and Sustainable Energy Reviews 108, 209—237.

Marsden, G., Frick, K.T., May, A., Dand Deakin, E., 2011. How do cities approach policy innovation and policy learning? A study of 30 policies in Northern Europe and North America. Transport Policy 18 (3), 501—512.

McDonald, R., Weber, K., Padowski, J., Flörke, M., Schneider, C., Green, P., Gleeson, T., et al., 2014. Water on an urban planet: urbanization and the reach of urban water infrastructure. Global Environmental Change 27, 96—105.

Meng, F., Liu, G., Liang, S., Su, M., Yang, Z., 2019. Critical review of the energy-water-carbon nexus in cities. Energy 171, 1017—1032.

Novosel, T., Cosić, B., Krajačić, G., Duić, N., Pukšec, T., Mohsen, M., Ashhab, M., Ababneh, A., 2014. The influence of reverse osmosis desalination in a combination with pump storage on the penetration of wind and PV energy: a case study for Jordan. Energy 76, 73—81.

NYU Urban Expansion Program, 2019. Lincoln Institute of Land Policy, UN Habitat, Atlas of Urban Expansion. http://www.atlasofurbanexpansion.org/.

OECD, 2017. International Transport Forum, Shared Mobility Simulations for Helsinki. Case-Specific Policy Analysis Report.

Ram, M., Bogdanov, D., Aghahosseini, A., Gulagi, A., Oyewo, A.S., Child, M., Caldera, U., Sadovskaia, K., Farfan, J., Barbosa, L.S.N.S., Fasihi, M., Khalili, S., Dalheimer, B., Gruber, G., Traber, T., De Caluwe, F., Fell, H.-J., Breyer, C., 2019. Global Energy System

Based on 100% Renewable Energy —Power, Heat, Transport and Desalination Sectors. Study by Lappeenranta University of Technology and Energy Watch Group, Lappeenranta, Berlin.

Region Skåne, 2014. The Scandinavian 8 Million City (Den skandinaviska 8 miljonersstaden) Slutrapport.

Renewables 100 Policy Institute, 2019. Go 100% Renewable Energy. http://www.go100percent.org/cms/index.php?id=19.

Rogelj, J., Shindell, D., Jiang, K., Fifita, S., Forster, P., Ginzburg, V., Handa, C., et al., 2018. Mitigation pathways compatible with 1.5°C in the context of sustainable development. In: Masson-Delmotte, V., Zhai, P., Pörtner, H.O., Roberts, D., Skea, J., Shukla, P.R., Pirani, A. (Eds.), Global Warming of 1.5°C, an IPCC Special Report on the Impacts of Global Warming of 1.5°C above Pre-industrial Levels and Related Global Greenhouse Gas Emission Pathways, in the Context of Strengthening the Global Response to the Threat of Climate Change, Sustainable Development, and Efforts to Eradicate Poverty (in Press). https://www.ipcc.ch/sr15/.

SCIS, 2019. Celsius Site Gothenburg. https://smartcities-infosystem.eu/scis-projects/demo-sites/celsius-site-gothenburg.

SDEWES Centre, 2017. SDEWES Index. http://www.sdewes.org/sdewes_index.php.

Seto, K., Dhakal, S., Bigio, A., Blanco, H., Delgado, G., Dewar, D., et al., 2014. Human settlements, infrastructure and spatial planning. In: Edenhofer, O., Pichs-Madruga, R., Sokona, Y., Farahani, E., Kadner, S., Seyboth, K., et al. (Eds.), Climate Change 2014: Mitigation of Climate Change. Contribution of Working Group III to the Fifth Assessment Report of the Intergovernmental Panel on Climate Change. Cambridge University Press, Cambridge.

SSN, 2019. Updates from Scottish Government on the Climate Change Bill. https://sustainablescotlandnetwork.org/news/update-on-the-scottish-government-s-climate-change-bill.

Steffen, W., Rockström, J., Richardson, K., Lenton, T., Folke, C., Liverman, D., Summerhayes, C., et al., 2018. Trajectories of the Earth system in the anthropocene. Proceedings of the National Academy of Sciences 115 (33), 8252–8259.

Stockholm City Executive Office, 2016. Strategy for a Fossil-Fuel Free Stockholm by 2040.

Stockholm Exergi. 2019. https://www.opendistrictheating.com/about/open-district-heating-how-it-works/.

Sun, L., Wang, S., Liu, S., Yao, L., Luo, W., Shukla, A., 2018. A completive research on the feasibility and adaptation of shared transportation in mega-cities — a case study in Beijing. Applied Energy 230, 1014–1033.

Taubenböck, H., Esch, T., Felbier, A., Wiesner, M., Roth, A., Dech, S., 2012. Monitoring urbanization in mega cities from space. Remote Sensing of Environment 117, 162–176.

Tirado-Soto, M., Zamberlan, F., 2013. Networks of recyclable material waste-picker's cooperatives: an alternative for the solid waste management in the city of Rio de Janeiro. Waste Management 33 (4), 1004–1012.

UNEP, 2017. District Energy in Cities Paris Case Study.

United Nations, 2018. Department of Economic and Social Affairs, the World's Cities in 2018: Data Booklet (ST/ESA/SER.A/417).

United Nations Division for Sustainable Development, 1992. Agenda 21. United Nations Conference on Environment & Development, Rio de Janerio, Brazil.

Wang, S., Yang, K., Yuan, D., Yu, K., Su, Y., 2019. Temporal-spatial changes about the landscape pattern of water system and their relationship with food and energy in a mega city in China. Ecological Modelling 401, 75–84.

Ward, J., Sutton, P., Werner, A., Costanza, R., Mohr, S., Simmons, C., 2016. Is decoupling GDP growth from environmental impact possible? PLoS One 11 (10), e0164733.

WHO Ambient (Outdoor), 2017. Air Pollution in Cities Database 2016. http://www.who.int/phe/health_topics/outdoorair/databases/cities/en/.

Wu, Q., Ren, H., Gao, W., Weng, P., Ren, J., 2018. Coupling optimization of urban spatial structure and neighborhood-scale distributed energy systems. Energy 144, 472–481.

Yamagata, Y., Seya, H., 2013. Simulating a future smart city: an integrated land use-energy model. Applied Energy 112, 1466–1474.

Zhang, Y., Wu, J., Zhou, C., Zhang, Q., 2019. Installation planning in regional thermal power industry for emissions reduction based on an emissions inventory. International Journal of Environmental Research and Public Health 16 (6), 938.

Future needs from the SMC plans — looking at Jiaxing, China: 40 years' development from numbers

5

Wang Weiyi, PhD

Associate Professor, Jiaxing University, Zhejiang Province, People's Republic of China

Case 1

Body parts

I. Economic aggregate has grown by leaps and bounds, and its comprehensive strength is greatly enhanced.

II. Citizens' income keeps increasing, and people's livelihood has improved significantly.

III. Three adjustments have been made to the industrial structures, and industrial restructuring has been gradually optimized.

IV. Great achievements have been made in opening to the outside world, and the level of opening-up has been constantly improved.

V. The scale of investment has been expanding, and infrastructure has improved significantly.

VI. The overall planning of urban and rural areas has achieved remarkable results, and the coordinated development is at the top of the list.

VII. The market transformation and upgrading are obvious, and the new type of industry develops rapidly.

VIII. Scientific and technological innovation has been promoted significantly, and social undertakings have developed in an all-round way.

IX. The overall environmental quality tends to be better, and the effect of energy saving and consumption reduction is remarkable.

(anunique@aliyun.com) presented by Jiaxing Statistics Bureau

Case 2

Title: Promoting the construction of ecological Jiaxing in an all-round way (about 15,000 bytes)

(Xin Jiaxing, Vol. 402, Dec. 2018)

Sustainable Mega City Communities. https://doi.org/10.1016/B978-0-12-818793-7.00005-6

Body parts

I. Jiaxing Path of Water Environment Governance
II. Exploration on Prevention and Control of Air Pollution
III. Construction and Practice of Beautiful Jiaxing
IV. Practice and Innovation of Environmental Protection Mechanism
V. Building a Model City of Water Township in the South of the Yangtze River

Case 3

Title: Implementing the strategy of connecting with Shanghai and fully integrating into synergetic development of Yangtze River Delta (about 15,000 bytes)
(Jiaxing University)

Body parts

I. To strengthen traffic connection and improve infrastructure
II. To promote platform convergence and closer industrial collaborative development
III. To promote industrial convergence and deepen complementary and integrated development
IV. To speed up the integration of people's livelihood and share the sustainable development of public services
V. To implement element alignment and make talents, science and technologies integrate more frequently

Case 4

Title: "No litigation" is the best business environment (about 5000 bytes)
(Jiaxing Daily, Mar. 16, 2019)

Body parts

I. Practicing refined and management and focusing on "extinguishing" at the source
II. Establishing rules and regulations and ensuring conflicts and disputes are handled in time and properly
III. Emphasizing the cultural influence and embedding "Harmony" culture in the heart

Local government partnerships for innovation: the case of Logan, Australia

By Ricardo Martello who is the City Futures Manager at Logan City Council, responsible for leading the innovation and smart agenda for the City of Logan. With qualifications in Innovation and Entrepreneurship, Urban Development and Sustainability, and Natural Resource Economics, Ricardo has experience in the private and public sector in both Australia and overseas.

In Australia, urban local governments are striving to earn their stripes as "smart cities." However, innovative practices do not emerge simply because a council sets a vision and a series of ambitious and well-intentioned goals.

For many councils, the very processes that define local government can be the greatest inhibitors, and the road to innovation can become a long journey of slow change.

At Logan City Council in Queensland, we have taken a different approach to influencing change by harnessing the power of collaborative partnership. We have done this through focused initiatives involving community, universities and secondary schools, social innovators, change makers, entrepreneurs, and our own staff.

The City of Logan is home to 320,000 residents from 217 different cultural backgrounds, spread across 70 suburbs. The city's economy has traditionally been underpinned by manufacturing and retail. We are located between two major cities: Brisbane, the state capital, and the high profile tourist strip of the Gold Coast; and we compete with them for investment dollars.

In 2016, our council set a new vision for Logan to become an "innovative, dynamic city of the future." Our civic leaders wanted to take advantage of emerging technology and the digital revolution to drive a new wave of economic growth for our city, which has historically had pockets of high unemployment.

To support that vision, Council created a new functional area to drive innovation—the City Futures Branch—which then developed Logan's first City Futures Strategy. The strategy was an important first step, but the team understood the next steps would determine whether the strategy became a wish list sitting on the shelf, or a living, breathing document driving real change.

The answer, happily, is the latter. In less than 12 months since its adoption, the strategy has delivered three successful initiatives under our Partnerships for Innovation approach. The Partnership has now been recognized with a shortlisting in the Apolitical Cities of the Future Policy Competition 2019.

The Partnership for Innovation entries—showcasing CityStudio Logan, INNOV8Logan and the Internal Collaborative Network—are the only Australian entries in the impressive international shortlist. These three projects are changing the way the Council staff think and respond to social innovation and entrepreneurial opportunities.

City studios Logan partnering with universities and high schools

In July 2018, we became one of only two local governments in Australia to sign-up with leading Canadian program, CityStudio Vancouver, which brings together staff, students, academics, and the wider community to work on social innovation projects.

The licensing agreement meant that we could use the CityStudio model to allow students and residents to create social innovation projects in partnership with Council staff. The key to the model is to match social challenges with stakeholders and social innovators and provide space and resources for them to develop a solution.

The initial 4-month pilot involved 22 Queensland University of Technology students (architecture, landscape architecture, urban planning, and gaming innovation) working with Council staff to influence master plans and activation strategies for key precincts in the city. CityStudio Logan has since expanded its partnership to include the University of Queensland, Griffith University, and a cluster of 13 government secondary schools.

The latest project involves Council collaborating with a YMCA secondary vocational school to create a greater sense of place at Logan's weekly Global Food Village Markets. The outdoor fresh food markets celebrate Logan's rich cultural diversity. They are held in a part of the city that has experienced image challenges due to its street culture and historical perception of crime issues.

CityStudio Logan was tasked with helping prepare the markets for filming of an international documentary by celebrity chef Ainsley Harriott in February 2019, and—parallel with that—consider ongoing place-making initiatives to enhance the visitor experience.

Participating students were involved in onsite visits and in-depth research, and worked with Council's City Futures Branch to brainstorm, prototype, test, and deliver place-making solutions.

The result is a pop-up street lounge made from materials from the Logan Recycling Market including a couch, vertical garden, art gallery, and book nook. It is a place where people can relax between browsing the food stalls, mingle, and enjoy the atmosphere.

INNOV8 Logan—Logan City, Australia has a virtual innovation hub

http://innov8logan.com.au was born in June 2018, when a group of likeminded innovators came together to share beer and pizza in an industrial building, and to talk about making it easier for local startups to make their big ideas a reality.

By the end of the gathering, the Logan Entrepreneurs Group had formed, along with a plan to create an online hub designed specifically for Logan, where local startups could find information and connect with the support they need at all stages of their journey.

The Logan Entrepreneurs Group is a collection of representatives from Logan-based startups, social enterprises, businesses, and entrepreneurs, along with Griffith University, the Queensland Government, and us. We have a seat at the table with the group, as we understand governments' role in building sustainable ecosystems, and have also provided extensive in-kind support.

One of the key drivers was to tap into the community of experts who were mentoring startups in neighboring cities. Because that knowledge was leaving our city, local startups were following. The idea was to keep both within Logan's boundaries to strengthen Logan's entrepreneurial ecosystem and deliver benefits back to our local community.

Central to it all was an online hub where a virtual community could be created and Logan's entrepreneurial ecosystem could be easily accessed.

INNOV8Logan was launched in September 2018, to a crowd of 300 innovators, startups, entrepreneurs, and investors. Queensland Chief Entrepreneur Steve Baxter was key note speaker, and attendees had the chance to see first-hand some of the cutting edge products being developed and manufactured in Logan. This included 9D virtual reality machines, plug and play swimming pools, batteries powered by recycled material that provide electricity to remote communities, solar-powered flood warning signs, and a wireless waterways system that monitors water quality in real time.

Six months on and INNOV8 Logan now supports Logan's entire innovation ecosystem. It has become a central point for the community to access innovation news, events, member profiles and stories, education, opportunities, connections, and opportunities to collaborate on projects. We are continuing to provide extensive in-kind support for INNOV8Logan, including hosting the online platform.

Council's internal collaborative network: a new mindset for staff

The third piece of the innovative partnerships picture is our own Internal Collaborative Network (ICN), created in July 2018 to empower Council staff to be a part of the organization's innovation agenda. The aim was to break down silos and activate innovative thinking through business improvement.

Interested staff from across Council's 1700-strong workforce self-nominated to be involved, and then participated in a 2-day facilitated workshop to scope and cocreate the ICN's purpose and charter.

From there, staff who mostly had not worked together before formed into four ICN teams. These teams were tasked with developing initiatives that addressed challenges associated with economic development, environment, efficiencies, and/or connectivity. Given great freedom and loose parameters, some teams thrived, while others found the lack of structure less effective.

The teams then pitched their solutions to Council's Executive Leadership Team. The winning pitch involved developing a behavior change campaign to encourage Council's large workforce to buy locally, particularly those staff who do not live locally and would otherwise drive in and out of the city each day without contributing to the local economy.

An internal "buy local" marketing campaign ran during the 12 days of Christmas, with staff encouraged to post their local shopping/dining experiences to social media and report on their dollar spend. As a result, at least $18,000 was spent locally by staff; a result that exceeded expectations. The initiative not only provided a local economic boost, it also helped change staff behavior in terms of where and how they spend money on food, goods, and services.

ICN participants across all challenges gained skills in defining and validating assumed challenges, building a solution using divergent thinking, testing that solution with customers, and engaging early adopters. ICN challenges unable to find traction were those with assumed challenges that could not be validated. These cases provided lessons in not being solution focused, but rather in being sure to first fully define the problem.

So, yes, local government innovation is possible. We just need to think in new ways and allow ourselves to be influenced through collaborations and partnerships.

What's next

- **CityStudio**: We are strengthening our university and secondary school partnerships, with the matchmaking process set to occur again in March 2019.
- **INNOV8Logan**: The online hub will help drive the focus and activities of Council's $6m Innovation Hub, due to open by early 2019.
- **ICN**: Staff can self-nominate to the network, with current participants closing the loop on existing projects in March and the next round of challenges to be set in coming months.

China

Greater Bay Area plan receives mixed reaction in Guangdong Province, still work to do for Beijing

China Economy (February 21, 2019)

- Hi-tech manufacturers, property speculators, and market investors who stand to gain support the plan for China's rival to Silicon Valley and the Tokyo Bay Area
- Export traders and low-tech manufacturers see Beijing's grand plan as doing little to help their struggle to survive, with fears it could actually increase costs

The Bay Area scheme seeks to join nine developed cities in the province's Pearl River Delta with Hong Kong and Macau into an integrated economic and business hub. Photo: AFP.

The Greater Bay Area Hong Kong development plan

- The bay area should be driven by innovation and led by reform
- Hong Kong, Macau, Guangzhou, and Shenzhen are the four core cities of the 11

- Governments in Hong Kong, Macau, and Guangdong should enhance communication and cooperate with mutual respect
- Authorities are to draw up plans to control financial risks and crack down on illegal activities

Key development areas

1. An international innovation and technology hub
   - More opportunities and better conditions for Hong Kong and Macau youths to start businesses in the bay area
   - Allow Hong Kong and Macau research and design firms in Guangdong to have the same treatment as mainland companies and benefit from national and provincial policies
2. Expediting infrastructural connectivity
   - Adopt new models of clearance procedures on the express rail and Hong Kong-Zhuhai-Macau Bridge
   - Improve electricity transmission between mainland and Hong Kong
   - Increase capacity at immigration ports in the bay area for a more efficient flow of people and goods
3. Deepen ties between Hong Kong and mainland financial systems
   - Stock connections between Shanghai and Hong Kong, as well as between Shenzhen and Hong Kong, will be enhanced
   - Eligible Hong Kong and Macau banks and insurance firms will be supported in opening branches in Shenzhen, Guangzhou, and Zhuhai
4. Quality of life, work, and travel
   - Encourage Chinese nationals in Hong Kong and Macau to work at state enterprises and agencies
   - Hong Kong and Macau residents working on the mainland could get the same rights to education, medical care, elderly care, housing, and transport as mainland residents
5. Education in the Bay Area
   - Consider letting Hong Kong teachers work in Guangdong
   - Children of Hongkongers working on the mainland could get same right to education as mainland residents
   - Higher-education institutions from different cities will be supported in running schools and programmes jointly
6. Cooperation between Guangdong, Hong Kong, and Macau
   - The establishment of a Greater Bay Area international commercial bank in Guangdong
   - Create an international and market-oriented business environment based on rule of law, under the jurisdiction and legal framework of mainland China
7. Ecological conservation
   - Strengthen water and air pollution control in the Pearl Delta River area

The introduction of smart green city and its prospect in China: case of Beijing as an example and Singapore as a benchmark

Yueqi Zhou

EMBA Graduate, Cheung Kong Graduate School of Business, Beijing, China

Introduction
Chinese cities need a change

China is entering a new stage of urbanization. In accordance to the *Statistical Communique of the People's Republic of China on National Economic and Social Development* in 2017, the current urbanization rate in China is 58.52%, a rocketed increase from 17% within only 40 years, and is still in its rise.

Admittedly, the speedy urban sprawl benefits from the rapid growth of Chinese economy. However, when the miracle of China's double-digit GDP growth ceased to be, Chinese government, enterprises, and even individuals, find themselves at a crossroad. Regarding the urbanization pattern, China has no good experience to learn from. After the excessive sprawl, Chinese people find that urban life seems not to be as convenient and healthy as it should be. This time, solely relying on the immethodical urban planning model, which is basically isolated between different domains, can never support the meteoric extension of cities. Our city planners have the commitment to deal with a more profound problem other than economic growth, city infrastructure, social welfare, or environmental protection exclusively, a problem concerning how to integrate available resources and obtain the maximization of utility in every social aspect. Smart and Green, that is the solution to the conundrum.

The introduction of smart green city

Smart city employs cutting-edge techniques such as Internet of things and cloud computing to collect and process massive amount of data and then optimize the city planning. For instance, issue of urban layout, public service, resource scheduling, and environment monitoring can all refer to the analysis result from the smart information network.

Urban layout

Land use efficiency of different functional area can be measured by the building-density, population density, and the output capacity. Smart information network can integrate information flows through modeling analysis and give out optimized schemes for land reallocation.

Public service

All citizens and public facilities are connected to the smart network through the urban net and the Internet of things. Specific information about roads that are in frequent congestion, public facilities that are obsolete, hospitals that are always crowded, and places that have high criminal rate can be sent by facilities access the Internet of thing or by citizens access the urban net. The cloud computing center then tracks the dynamic changes in that region in real time and gives feed back to the information receivers such as government and citizens. With those detailed and accurate references regarding traffic situation, infrastructure maintenance, public health service, and public security, government can make more targeted municipal decisions, and citizens can enjoy guaranteed life.

Environment monitoring

Smart environment monitoring system monitors the key ecological index and gives quick response to the abnormal fluctuations by sending reminder to nearby citizens, municipal government, and the responsible party. Such systems can also provide the most economical plan for sewerage treatment and garbage collection. The environment and resource management is not necessary to be limited within a city since the smart network between cities can cooperate and formulate a joint monitoring scheme that may greatly improve the overall utility of the environmental protection facilities.

   Green city, based on the smart network and macro urban planning, minimizes the overall carbon and harmful gases emission and maximizes the resource utilization efficiency in order to create the virtuous cycle between urban development and ecological protection. Feasible approaches include green space and green building construction, energy-related policies and regulations, and management of trash recycling and sewerage.

Green space and building

Green space can be allocated according to the building density. For residential area with sufficient space between buildings, large green space such as park can be set up to improve the habitability of the neighborhood. As for CBD where building density is too high to set up ground green space, alternatives like roof gardens and plant walls are also helpful in reducing the heat island effect.

Energy-related policy and planning

Though the large urban population contributes to the unimaginable amount of energy consumption, it can also be a powerful driving force in the promotion of new energy and energy conservation. Government can guide a huge transformation of energy consumption structure by enacting related regulations and delivering smart urban planning. First, government can support new energy industries with specific fiscal policies and encourage citizens to purchase new energy products through subsidy. Besides, energy conservation should be an orientation in city planning. For example, more bicycle greenways and bus routes mean less carbon emission per capita. A mixed-use community that can satisfy residents' daily demand without making them drive elsewhere can be very helpful in reducing the pollution caused by traffic.

Garbage classification and sewage treatment
Precision garbage classification can greatly increase the recyclability of resources and reduce the environmental burden caused by waste incineration and landfill. Since infiltration of poor-treated sewage may cause serious chain effect through the circulation of underground water, advanced sewage treatment facilities are therefore indispensable for ensuring the ecological balance and safety.

The method to measure the efficiency of urban planning

Introduction of Data Envelopment Analysis model

Data Envelopment Analysis (DEA) is a nonparametric method of evaluating the relative effectiveness of the decision-making unit (DMU) based on multiple input and output indicators employing the linear programming method. DEA was first proposed by Charnes, Cooper, and Rhodes (CCR) in 1978. The CCR model can measure whether production is technically effective and scale effective, based on constant returns to scale. In 1984, Banker proposed another DEA model (BCC) which can evaluate the relative effectiveness of production with variable returns to scale. CCR and the BCC models are the most classic and widely used models among the DEA models.

The decision-making unit with the best input-output efficiency constructs the production frontier, which can enclose all DMU in the production front. The unit on the frontier surface is called the DEA effective unit, and the remaining enveloped units are noneffective units. By calculating the deviation of the noneffective unit from the production frontier, efficiency rate can be evaluated.

The weight of multi-inputs can be written as a vector: $\omega = (\omega_1, \omega_2, ..., \omega_m)^T$
The weight of multi-outputs can be written as a vector: $\mu = (\mu_1, \mu_2, ..., \mu_m)^T$
The efficiency evaluation of the kth DMU can be presented as:

$$\begin{cases} \max \dfrac{\mu^T y_k}{\omega^T x_k} \\ \omega^T x_j - \mu^T y_j \geq 0 \quad j = 1, ..., n \\ \omega \geq 0, \mu \geq 0 \end{cases}$$

Traditionally, to set up a function, the analyst needs to preset coefficient for each independent variable, making the evaluation biased. It is also difficult to perceive the relationship between these factors by using specific functions.

However, when using DEA, explicit functional relationship is not required. DEA has the following advantages: First, in DEA, the optimal efficiency of DMUs is unrelated with the dimensionality of the input and output index, making the nondimensionalization unnecessary. Second, DEA is very objective as the algorithm will automatically assign weight to each input indicator to optimize the evaluation of dependent variables. Finally, DEA assumes that each input indicator is associated with one or more outputs, thus no explicit equation is needed. Therefore, since the core concept of smart green city is to make the most of every input factor and obtain the highest outcome in the whole society, DEA is a suitable model to employ when we try to asset the overall urban planning efficiency.

Sample cities selection

Two cities are selected as the research samples: Beijing and Singapore.

Beijing is the political center, economic center, and the cultural center of China. It enjoys China's greatest financial investment and most preferential policy, has the most typical industrial structure that dominated by the secondary and tertiary industries, while also troubled by the most common urban disease. In a word, to analyze Beijing is to analyze the other hundreds of cities in China.

Singapore is a representative and a successful example of smart green city. In 2006, Singapore enacted a 10-year plan named Smart Nation, aiming at constructing itself into an intelligent country and a globalized city. Thus, taking Singapore as a benchmark can partially measure the growth potential of Beijing if it ultimately transforms into a smart green city.

Indicators selection

For the input side, three indicators are selected: the total investment in secondary and tertiary industries, the employment figure in secondary and tertiary industries, and the annual energy consumption. (The indicators rule out the primary industry part since Beijing and Singapore are both dominated by the tertiary industry—dominant cities while the primary industry, which is mainly urban agriculture, can be regarded as an insignificant component.) The input indicators measure three typical means of production: the total investment represents the capital input; the employment figure shows the labor input; and the energy consumption reflects the energy input.

For the output side, four indicators are selected: regional/national GDP (though the value of agricultural production is included in GDP, it only accounts for less than 1% in both Beijing's GDP and Singapore's GDP, thus is negligible), registered health personnel number, total paved roads length, and the concentration of PM2.5. To evaluate the overall efficiency of a city's planning, factors concerning different aspects, including economic, social, and ecological aspects, should be taken into account. The regional/national GDP serves as a reference of economic benefit; the registered health personnel number and the total paved roads length indicate the social benefit; the concentration of PM2.5 embodies the negative ecological benefit.

The statistics resources

Statistics of Beijing are from Beijing Municipal Bureau of Statistic.

Statistics of Singapore are from Department of Statistics Singapore.

Data processing statement:

The USD-RMB exchange rate is taken as 1:6.9663.

All numbers are rounded down to integral numbers.

DEA result analysis

By employing the CCR and BCC model in DEA, the technical efficiency (TE), the pure technical efficiency (PTE), the scale efficiency (SE), and the returns to scale (RS) of the two regions can be compared.

The comparison of technical efficiencies between Beijing and Singapore

Applied model: DEA-CCR-TE

Technical efficiency (TE), based on the assumption of constant returns to scale, measures the distance between the current input and output position of the DMU and the production frontier, reflecting both the technical efficiency and scale efficiency.

The TE of Beijing declines almost every year from 1 in 2008 to 0.704 in 2017, losing almost 30% of its technical efficiency. Singapore stays in a constant high position of nearly 1 with only a slight decline in 2014, indicating that the overall planning in Singapore, including the input structure (capital, labor, resource) and output (economics benefit, social benefit, ecological benefit) is highly efficient and sustainable.

The comparison of pure technical efficiency between Beijing and Singapore

Applied model: DEA-BCC-PTE

Pure technical efficiency (PTE), based on the assumption of variable returns to scale, evaluates the distance between the current input and output position of the DMU and the production frontier. PTE normally measures the efficiency of flexibility index such as innovation and management.

The pure technical efficiency of Beijing and Singapore is relatively close. And in 2014, the PTE of Beijing exceeded that of Singapore, indicating that Beijing has a comprehensive and mature management system regarding the input and output resources.

The comparison of scale efficiencies between Beijing and Singapore

Applied model: DEA-BCC-SE

The scale efficiency (SE) is obtained by dividing the technical efficiency and pure technical efficiency (SE = TE/PTE). It examines the deviation between the current scale of production and the optimal scale of production of the DMUs.

The trend of scale efficiency (SE) of Beijing and Singapore's urban planning is almost identical with the technological efficiency. Scale efficiency of Singapore's urban planning stays in a constantly high level, reflecting that Singapore stays moderate in the input side so as to maintain the high marginal output, while Beijing undergoes a continuous declination in its scale efficiency owing to the deviation from the best input structure. Therefore, Beijing ought to optimize the proportion of the input components by reconsidering its urban planning in order to obtain a higher city output without wasting too many resources.

Conclusion

The DEA result is worth thinking about. Singapore's total area is one-twelfth that of Beijing, while its population density is six times that of Beijing. But its efficiency of urban planning is much higher. Thus, to be objective, the input resources are not the restrictive factors that hinder the further development in Beijing. It is the lack of smart urban planning and the overlook of the ecological value that leads to the undesirable output. However, the good news is Beijing still has considerable growth potential even without increasing the input volume, so long as the marginal output can be improved. With reference to Singapore's construction achievements, China's cities can be expected to bring about the driving force for national development and the increase in people's welfare in the future.

The prospect of smart green city in China
Smart green agglomeration

There are still large underdeveloped and vacant suburban areas around Chinese cities, enabling the possibility to develop agglomeration. When some mega cities plan to simplify their urban function and transfer some industries and institutions to the surrounding areas, those suburban areas will undoubtedly play a vital role. The future smart green cities in China are very likely to be in the form of agglomeration leading by a central city, with satellite cities surrounding. In the smart green agglomeration, the central city assumes the core functions of the entire group while the satellite cities undertake specific functions utilizing the industrial agglomeration effect and the scale benefit.

The driving force from advanced Mobile Internet industry

China's Mobile Internet technology is incredibly advanced, and it has the world's largest mobile Internet user group. Such can be a prerequisite for the prosperity of smart green cities.

The mobile Internet active users in China reached 970 million by the end of 2017. The world's largest mobile Internet user group serves as a solid foundation for the usage of smart network in the future. Along with that, China's leading-edge 5G network technology will realize the connection between Internet of Things, AI, and mobile Internet. And the booming digital economy greatly popularized the mobile payment technology. Therefore, the high mobile Internet penetration rate and the well-developed related industries can be a good foundation for the smart green urbanization.

The incentive from the political achievement evaluation system

In August 2014, eight ministries and commissions in China collectively issued the *Guideline for promoting the healthy development of smart cities*, the first guideline that was ever published concerning the comprehensive construction of smart cities. According to the guidelines, by 2020, China will complete the construction of the first batch of smart cities which are intensive, intelligent, green, and low-carbon.

The political achievement evaluation system is a standard to evaluate the governing capacity of a local government. Basically, the standard of this system is closely related to the local GDP. However, with the guideline issued by the central government, the political achievement evaluation system may also attach greater importance on the smart and green urban planning.

When more quantizable indicators like the carbon emission, green space coverage, and the renewable energy usage rate are taken into account, local governments will have increased incentives to push forward the smart green urbanization.

By and large, it is high time for Beijing, and the other cities in China and even around the world, to change the thinking pattern of urban planning and embrace the promising future of smart green city. Though it will be a long way to realize the vision that everyone can live in a livable and sustainable city, every small step we take today will lead us closer to the brave new world.

Further reading

Beijing, 28 10 2018. Municipal Bureau of Statistics. http://www.bjstats.gov.cn/English/.
Singstat, 18 10 2018. https://www.singstat.gov.sg.
Banker, R.D., Chang, H., Cooper, W.W., 1996. Equivalence and implementation of alternative methods for determining returns to scale in data envelopment analysis. European Journal of Operational Research 473−481.

Future needs from the Smart Mega City (SMC) plans—smart green city—the case of Istanbul

6

Sevinç Gülseçen, PhD[1], Murat Gezer, PhD[2], Serra Çelik, PhD[2], Fatma Önay Koçoğlu, PhD[2]

[1]*Chair and Director, Informatics Department, Istanbul University, Computer Science and Application Center, Istanbul, Tukey;* [2]*Professor, Informatics Department, Istanbul University, Istanbul, Turkey*

Overview

In recent years, the concept of the smart city has influenced urbanization strategies around the world, regardless of whether the cities are small or large (Caragliu et al., 2011). Increasing speed of citizens' use of mobile devices and the internet has increased the interest in this topic in recent years, ranging from pollution and energy consumption of large cities to environmental safety (Oberti and Pavesi, 2013).

There is not yet a universally accepted definition, with too many definitions of Smart City. A major obstacle in defining this term is the uncertainty of the meanings attributed to the word "smart" and the label "Smart City." Smart City "often links technological information transformations with economic, political, and sociocultural change" (Hollands, 2008). Cocchia (2014) explains the smart city as a city based on the "smart" knowledge city, ubiquitous city, sustainable city, digital city, etc. Smart cities and digital city definitions are most frequently used to show the wisdom of a city, that a city can be intelligent and digital, use technology to improve the quality of life in the city, especially Information and Communication Technology (ICT) but this idea is somewhat outdated (Tokmakoff and Billington, 1994).

Smart city

A smart city is a complex result with several dimensions. These dimensions are technology, citizens, public and private bodies, and urban vision (Pardo and Taewoo, 2011). In addition, all cities in the world have differences in cultural, economic, and social dimensions. Every city wants to apply common smart city idea and to follow their own specific goals. According to Giffinger et al. (2007), "A Smart City is a city well performing built on the 'smart' combination of endowments and activities of self-decisive, independent and aware citizens."

Sustainable Mega City Communities. **https://doi.org/10.1016/B978-0-12-818793-7.00006-8**

Pardo and Taewoo (2011) divided the most commonly used terms into dimensions according to their common characteristics. These dimensions are as follows:

1. Technology dimension: it is based on the use of infrastructures (especially ICT) to improve and transform life and work within a city in a relevant way. This dimension includes the concepts about digital city, virtual city, information city, wired city, ubiquitous city, and intelligent city;
2. Human dimension: it is based on people, education, learning, and knowledge because they are key drivers for the smart city. This dimension includes the concepts about learning city and knowledge city
3. Institutional dimension: it is based on governance and policy, because the cooperation between stakeholders and institutional governments is very important to design and implement smart city initiatives. This dimension may include the concepts about smart community, sustainable city, and green city.

Sometimes a smart city project can be seen as a drug that can solve all urban problems such as environmental pollution, local transportation difficulties, and economic crisis. These expectations are not supported by an open intelligence vision of the city or an effective intelligent program and initiatives (Mulligan and Olsson, 2013). Some cities are able to define themselves as smart cities without a strategic vision of the future. The lack of a smart strategic vision negatively affects the performance of smart projects and initiatives (Kourtit et al., 2013).

The vision of a smart city depends on the importance of technology, especially ICT. It is due to its strong need to solve the problems that affect life in large metropolises such as digital city idea as well as traffic, pollution, energy consumption, waste treatment, and water quality. These directions also bring to mind the green city. Environmental themes are an important part of intelligent city goals (Dameri and Rosenthal-Sabroux, 2014). In this smart city vision, initiatives focus on improving city wisdom:

- Reducing energy production, energy cost, and CO_2 emissions from renewable sources and balancing increased energy demand in cities
- Reducing effective energy demand and consumption
- Local transport quality and greenery, reduce transportation pollution

Smart green city

All over the world, especially with the development of the industry, population density has begun to gather in the cities and urbanization has accelerated. Urbanization has brought about various problems in transportation, housing, health, infrastructure, etc. Above all, problems are increasing exponentially due to the increasing depletion of natural resources and increasing carbon atmospheres because of usage of nonenvironmentally friendly solutions and nonconvertible wastes (URL1). Although human beings continue their existence in living spaces that are gradually getting away from nature, nature is on the way to disappearing day by day. Countries are trying to provide solutions for these problems, which pose a huge threat in the global sense, and to offer better life opportunities for their citizens. Especially developed information and communication technologies are used for the solutions, the data collected by various technological tools are converted to knowledge, and intelligent applications are being developed. In this respect, the concept of "Smart City" has become increasingly widespread and various projects have been passed on to international platforms. "Smart City" applications, such as city guides transferred to web portals, e-services offered to citizens (Anthopoulos, 2015), and various transportation solutions, facilitate the lives of city people. However, in addition to these applications, issues such as security/safety, transportation, and sustainable and efficient energy are also referred to with the concept of "Smart City" (Lombardi et al., 2012). One of these issues is "green." As implementation of the smarter applications on transport, energy, and city planning can be over costing for more polluted cities, the concept of "green city" is becoming more important (Neirotti et al., 2014). When technology-based solutions that reintegrate human life with green are taken into consideration, the concept of "Smart City" evolves into the concept of "Smart Green City."

To be considered as a green city, various criteria can be listed in water and air quality, green area distributions, and waste management. Zygiaris (2013) stated that one of the six key factors in the creation of sustainable and smart cities is the green environment. He also expressed what is needed for the green environment in the form of planning of alternative energy sources, conservation of water resources, green building policies, green transportation policies, and master plans for CO_2 reduction. Su et al. (2011) pointed out the importance of interoperability of systems built with different devices for green city construction and stated that more efficient use of resources can be achieved through various monitoring and warning systems. Similarly, Kim et al. (2012) also proposed a model for becoming a smart green city with an integrated energy monitoring and visualization system based on large data management using a web-based platform.

As a result of the work carried out by Economist Intelligence Unit (EIU) and sponsored by Siemens, a green city index was established according to more than 120 countries included in the survey and five different regions created. Strategies for CO_2 intensity, emission, and reduction; general energy consumption, renewable energy production, and various energy policies; the state of energy consumption of buildings; green transportation measures and measures to reduce traffic intensity; waste management, waste recycling, and waste generation; the use of water resources; air quality and environmental management have been considered in the creation of the green city index (EIU, 2012). While defining and describing the concept of today's green city in this way, Jedliński (2014) stated that in the future intelligent green cities will have features such as structure deindustrialization, greater heterogeneity, holistic thinking, functionality maximization, greater humanization, lower power supply cost, and increase of the public goods utility. The following applications are crucial for the construction of green cities:

"Renewable Energy Resources" instead of fossil fuels to reduce energy consumption and carbon emissions:

- Smart applications (building, transportation, waste management, etc.)
- Precautionary actions for climate change (technology based)

- Smart metering systems
- Integrated systems
- Smart grids

To summarize, the use of energy-efficient smart buildings, increased green areas for carbon dioxide capture, improved traffic measures to reduce CO_2 emissions, effective waste management, the use of applications to improve water and air quality, and a green lifestyle that facilitates access to products and services will create a system in which carbon footprint is reduced and sustainability is effective (URL2), and this will enable "smart green cities."

Istanbul green smart city projects

Istanbul is the only city in the world that is located on two continents: Europe and Asia, and according to UN, The World Cities report, it is the 15th most populated city in the world, where 15 million people are living (URL3) (Fig. 6.1)

Similar to other cities in the world, green smart city projects are being developed and used in Istanbul as well. As shown in Figure 6.1, these projects are ranged from climate change to conservation of green spaces, to intelligent transportation solutions, to energy-saving solutions, etc. Leading role in the implementation of the vision of a smart city is Istanbul Metropolitan Municipality that established an office and leads this vision. Some of the example projects that have been developed in the last 5 years are summarized in Table 6.1. From 1980s onwards, the Istanbul Metropolitan Municipality has been deceiving itself with its various subunits (ISBAK Istanbul IT and Smart City Technologies Inc., IETT, METRO Istanbul, etc.) of smart green city technology.

FIGURE 6.1

Green smart city initiatives.

Table 6.1 Smart green projects in Istanbul.

Project number	Project name	Project type	Project description
1	IBB WIFI	DC—communication	The Free Internet Project from Istanbul Metropolitan Municipality opened for the first time on April 1, 2014 at 27 locations (URL4). Every citizen have daily 1 GB Quota with 2 Mbps speed free for use
2	Mobiett/IETT	SC—smart transport and mobility	11,569 bus stops in Istanbul were made intelligent by using sensors. The approaching times of the buses can be seen from the screens in the bus stops. It is also seen through the IOS or Android application within the mobile phones (URL5)
3	Smart Bus/IETT	SC—smart transport and mobility	Free internet and phone chargers for the use of the passengers in the city buses have been added (URL6)
4	IBB Navi/İSBAK	SC—smart transport and mobility	Smart city navigation "İBB Navi" application was developed attempting to manage the urban traffic efficiently (URL7)
5	Traffic Density Map/İSBAK	SC—smart transport and mobility	With this application, it is aimed to inform users real time and make them to be oriented for alternative routes so that the demand to traffic-dense regions can be decreased and road network capacity can be used efficiently (URL7)
6	Smart City Project Office	No-tech	Smart City Project Office is established to conduct sustainable works regarding national/international resources in cooperation with the global companies experienced in implementation of similar smart city projects
7	Mahallem Istanbul	DC—data/ information/service	It will enable to obtain information via web or mobile application about urban/public services at the neighborhood level in Istanbul (URL8)

Continued

Table 6.1 Smart green projects in Istanbul.—*cont'd*

Project number	Project name	Project type	Project description
8	Smart City Best Practices Workshop	No tech	İstanbul Metropolitan Municipality (IMM) continues studies to apply "Smart City" concept in İstanbul that is one of the biggest metropolises of the world (URL9)
9	ISBAK	SC—company	ISBAK Istanbul IT and Smart City Technologies Inc. has been established by the Istanbul Metropolitan Municipality in year 1986 with the purpose of providing project design and implementation services through traffic and system engineering. Nowadays its mission is transfer their know-how to Smart City Projects (URL9)
10	İklim Istanbul	GC—climate change prevention	Works for reduction of greenhouse gases causing climate change and adaptation of the city to climate change (URL10)
11	Calculation of Personal Carbon Footprint	GC—climate change prevention	A web site about Calculation of Personal Carbon Footprint (URL11)
12	SEA SOLAR POWER PLANT Project	SC/GC/energy	It provides 240 kW of energy and 202 house use this energy per year. The target of this project is that 55% of the energy used in Istanbul until 2033 will be renewable (URL12)
13	BluesCities	SC	This Horizon 2020 project focus on the need to integrate water and waste within the smart city approach (URL13)
14	Neighborhood markets	GC/farm fruit and vegetables close to consumer	Istanbul is home to many neighborhood markets. These markets are established in certain days of the week.
15	Historical Vegetable Gardens of Yedikule	GC/urban farming projects	Yedikule is a neighborhood of old city Istanbul. According the literature urban agriculture started before 1500 years and still continue

Table 6.1 Smart green projects in Istanbul.—*cont'd*

Project number	Project name	Project type	Project description
16	G-Charge	SC/energy efficiency	Electric vehicle charging station in pilot areas in ISPARK parking lots (URL14)
17	Solar G-Charge	SC/energy efficiency	Use of solar energy at parking stations to improve energy efficiency and sustainable living
18	CitySDK	SC	Istanbul is part of the CitySDK (Smart City Service Development Kit and its application Pilots) project, which is funded by the European Union's ICT Policy Support Program as part of the Competitiveness and Innovation Framework Program (CIP) and led by Forum Virium Helsinki (URL15)
19	Smart Cities Technology Center	SC/DC	IBM launched Turkish Smart Cities Technology Center, which provides technological solutions to the problems generated by urban life (URL15)
20	Istanbul Airport	SC/DC	Istanbul Airport uses Internet of Things, mobile applications, face recognition systems technologies for smart passenger systems (URL16)
21	Istanbul Smart Grids and Cities Congress and Fair	SC/GC/DC	ICSG is an international congress that seeks to bring researchers, practitioners, developers, and users together to explore and discuss Smart Cities Technologies and Policies. (URL 18)
22	World Cities Congress Istanbul	SC/GC/DC	World Cities Congress aims to bring together leading institutions and organizations working on smart cities (URL 19).

SC: Smart City, GC: Green City, DC: Digital City.

Istanbul: not only smart but also a creative city

Smart cities represent an innovative way to achieve better interaction between the city stakeholders. The citizen can also interact directly with the government and hence contribute to the development and renewal of ideas to the benefit of the community. ICT players have an essential role here: they support the whole infrastructure and platform of applications while ubiquitously connecting the stakeholders in the city.

Istanbul was the capital as well as a center for trade and distribution from the mid-16th century till the beginning of the 20th century. With its huge population, Istanbul has hosted for ages communities from different civilizations and cultures who lived in harmony. There are approximately 400 cities in the world, each of which has more than 1 million citizens. The largest metropolitan area, Tokyo, has 28 million citizens; New York City has over 20 million. According to the 2004 United Nations HABITAT report, 60% of the world's population will live in a city by 2030.

Within the framework of the implementation of the United Nations 2030Agenda for Sustainable Development and the New Urban Agenda, the UNESCO Creative Cities Network provides a platform for cities to demonstrate culture's role as an enabler for building sustainable cities (URL17).

With its characteristics as a smart green city, in 2018, Istanbul has been selected as one of the "CITY of DESIGN" within the UNESCO Creative Cities Network.

Conclusion

In accordance with the purpose of creating smart green cities, various studies, such as intelligent transportation, intelligent energy, sustainability, and particularly the priorities related to the city needs, have been carried out. Furthermore, some issues such as budget requirements, IT opportunities, IT managerial capability, personnel requirement, and prejudice against information security should be overcome to perform these studies. That is to say, this subject should be handled from an economic and social perspective. Therefore, it is obvious that it needs time to structure smart green cities worldwide commonly.

Istanbul, one of the important centers of the world is also one of the cities that pays attention to creating smart green cities and tries to catch the trend of smart green cities considering the studies made and carried out successful applications especially in the fields of transportation and energy with both public and private sector investments. It is clear that Istanbul will continue to sustain this success with regards to the intended projects. On the other hand, Istanbul is home to both national and international events. In this way, smart green cities subject is discussed, and the exchange of ideas and various international collaborations are provided. Although Istanbul plays an active role in international platforms with the events such as International Istanbul Smart Grids and Cities Congress and Fair and World

Cities Congress. The city also handles the smart green cities subject, especially in the local administration with the events such as Smart Cities themed Istanbul Informatics Congress organized by The Informatics Association of Turkey.

References

Anthopoulos, L.G., 2015. Understwanding the smart city domain: a literature review. ISBN: 978-3-319-03166-8. In: Transforming City Governments for Successful Smart Cities. Springer-Cham, Switzerland.

Caragliu, A., Del Bo, C., Nijkamp, P., 2011. Smart cities in Europe. Journal of Urban Technology 18 (2), 65−82. Routledge.

Cocchia, A., 2014. Smart and digital city: a Systematic literature review. In: Dameri, R.P., Rosenthal-Sabroux, C. (Eds.), Smart City How to Create Public and Economic Value with High Technology in Urban Space. Springer, pp. 13−43.

Conference on Electronics, Communications and Control (ICECC), 9-11 September 2011, China. pp. 1028−1031.

Dameri, R.P., Rosenthal-Sabroux, C., 2014. In: Dameri, R.P., Rosenthal-Sabroux, C. (Eds.), Smart City and Value Creation, Smart City How to Create Public and Economic Value with High Technology in Urban Space. Springer, pp. 1−12.

Economist Intelligence Unit (EIU), 2012. The Green City Index. https://www.siemens.com/entry/cc/features/greencityindex_international/all/en/pdf/gci_report_summary.pdf.

Giffinger, R., Fertner, C., Kramar, H., Kalasek, R., Pichler-Milanovi, N., Meijers, E., 2007. Smart Cities: Ranking of European Medium-Sized Cities. Centre of Regional Science (SRF), Vienna University of Technology, Vienna, Austria from. http://www.smart-cities.eu/download/smart_cities_final_report.pdf.

Hollands, R.G., 2008. Will the real smart city please stand up? City: Analysis of Urban Trend, Culture, Theory, Policy, Action 12 (3), 303−320.

Jedliński, M., 2014. The position of green logistics in sustainable development of a smart green city. Procedia-Social and Behavioral Sciences 151, 102−111.

Kim, S.A., Shin, D., Choe, Y., Seibert, T., Walz, S.P., 2012. Integrated energy monitoring and visualization system for Smart Green City development: Designing a spatial information integrated energy monitoring model in the context of massive data management on a web based platform. Automation in Construction 22, 51−59.

Kourtit, K., et al., 2013. 11 an Advanced Triple Helix Network Framework for Smart Cities Performance. In: Smart Cities: Governing, Modelling and Analysing the Transition, 196.

Lombardi, P., Giordano, S., Farouh, H., Yousef, W., 2012. Modelling the smart city innovation. The European Journal of Social Science Research 25 (2), 137−149.

Mulligan, C.E.A., Olsson, M., 2013. Architectural implications of smart city business models: an evolutionary perspective. IEEE Communications Magazine 51, 6.

Neirotti, P., De Marco, A., Cagliano, A.C., Mangano, G., Scorrano, F., 2014. Current trends in Smart City initiatives: some stylised facts. Cities 38, 25−36.

Oberti, I., Pavesi, A.S., 2013. The triumph of the smart city. Techne: Journal of Technology for Architecture & Environment 5.

Pardo, T., Taewoo, N., 2011. Conceptualizing smart city with dimensions of technology, people, and institutions. In: Proceedings of the 12th Annual International Conference on Digital Government Research. ACM, New York, pp. 282−291.

Su, K., Li, J., Fu, H., 2011. Smart city and the applications. In: Proceedings International.

Tokmakoff, A., Billington, J., 1994. Consumer services in smart city Adelaide. In: Bjerg, Borreby (Eds.), Home-oriented Informatics, Telematics and Automation.

URL1. World Wildlife Fund- Turkey. Ekolojik Ayak Izi. https://www.wwf.org.tr/basin_bultenleri/raporlar/yaayan_gezegen_raporu/yasayangezegenrap.oru2014/ekolojikayakizi/.

URL2. 2018. Sürdürülebilir Şehirler. http://www.siemens.com.tr/web/1772-13670-1-1/siemens_turkiye_-_tr/siemens_turkiye/etkinlikler/kentlesme.

URL3. Turkish Statistical Institute. http://www.turkstat.gov.tr/PreTablo.do?alt_id=1059.

URL4. İBB Wifi Hizmetleri. http://isttelkom.istanbul/hizmetlerimiz/ibb-wifi-hizmetleri/.

URL5. Mobiett ile artık her durak akıllı. http://www.iett.istanbul/tr/main/news/mobiett-ile-artik-her-durak-akilli/1397.

URL6. Akıllı İETT Otobüsleri Artık Yollarda. http://www.iett.istanbul/tr/main/news/akilli-iett-otobusleri-artik-yollarda/1344.

URL7. Intelligent Transportation Systems. http://isbak.istanbul/intelligent-transportation-systems/.

URL8. M. Istanbul. http://www.mahallemistanbul.com/.

URL9. Smart City Best Practices Workshop Was Performed. http://isbak.istanbul/smart-city-best-practices-workshop-was-performed/.

URL10. İ. Istanbul. https://www.iklim.istanbul/.

URL11, https://www.iklim.istanbul/karbonhesap/.

URL12. ST E. Enerji, "Türkiyenin Ilk GES'i Eketrik Üretmeke Başladı. Ocak 2018, Sayfa 20. http://dergi.stdergileri.com/st-elektrik-enerji-2018-ocak/mobile/index.html.

URL13. reportBlueSCities Report Summary. https://cordis.europa.eu/result/rcn/190033_en.html.

URL14. GERSAN. İSPARK otoparklarını elektrikli araç şarj istasyonları ile donatıyor. http://www.g-charge.com.tr/statics/news/gersan__ispark_otoparklarini_elektrikli_arac_sarj_istasyonlari_ile_donatiyor/. UN The World's Cities in 2016 Report. http://www.un.org/en/development/desa/population/publications/pdf/urbanization/the_worlds_cities_in_2016_data_booklet.pdf.

URL 15. Smart Cities in Turkey. https://build.export.gov/build/groups/public/@eg_tr/documents/webcontent/eg_tr_092324.pdf.

URL 16. İstanbul New Airport set to be The New Flag-bearer of Internet of Things. http://www.igairport.com/en/press-center/press-room/istanbul-new-airport-set-to-be-the-new-flag-bearer-of-internet-of-things.

URL 17. UNESCO Yaratıcı Şehirler Ağı. http://www.designcityistanbul.com/Sitepage/index/14.

URL 18. International Istanbul Smart Grids and Cities Congress and Fair.

URL 19. World Cities Congress Istanbul.

Zygiaris, S., 2013. Smart city reference model: Assisting planners to conceptualize the building of smart city innovation ecosystems. Journal of the Knowledge Economy 4 (2), 217–231.

Economic options: back to the future: in China, the future is *now*

3

Woodrow W. Clark, II MA[3] PhD

Breaking news

News was made worldwide when China's CNOOC oil and gas company bid for American-owned UNOCAL. Unfortunately, parochial and narrow-minded American politicians decided that China should not acquire an American oil and gas production company. The same kind of discussion within the political arena also occurred over the sale of Maytag to a Chinese company.

In both cases, the politics was too much and the Chinese companies lost this round to becoming more international as well as addressing their future energy needs. The Economist (August 27, 2005) in late August ran a front page cover, titled "The oiloholics" which depicts both images of China and the United States drinking large quantities of oil and gas from barrels as the market prices hit over USA $70 per barrel which is three times the price from a year ago (Summer 2004). Costs to each country and its consumers are staggering. Indeed, there is competition for oil and gas supplies. Moreover, the two biggest world consumers will now drive those costs up even higher.

The American excuse or rational was that China was "economically invading" its domestic energy production. Clearly, according to some politicians, this is a threat to American security. Now where have we heard this before? Twenty years ago, in the 1980s, when the Japanese did something similar in buying overvalued American companies and real estate. We all know what happened there since then? The Japanese bubble burst. And the two counties are strong and fast economic and political allies today. Unfortunately, history repeats itself. Now it is China's turn to suffer this parochial worldview.

Energy is now on the China and US common agenda. And herein lies the opportunities for both countries. Sooner than later, each country, perhaps together—image that?—must develop renewable and sustainable energy supplies. Herein lies the economic future for both countries and the world for that matter. In a white paper, "California's Next Economy" (January 2003), the point was made that sustainable development was indeed one of California's new economic engines. And now today, that has indeed occurred in a range or areas including green public and private buildings (solar roof programs), use of Leadership in Energy and Environment Design Standards (LEEDS), California Hydrogen Highway, distributed energy generation, Fuel Cell Partnership and Collaboration, and much more.

Consider the tourists who are dining out in Shanghai's Bund with lighting for their needs and along the Bund itself. Here, the government orders office buildings to be lighted until 11 p.m. Meanwhile, factories go dark. The issue is not a choice of one energy use over another, but both can be accomplished.

Opportunities for economic growth: green development based upon sustainable businesses

While the actual numbers are hard to quantify, the trends can be seen in a variety of areas as energy needs and demands are met in the United States. California is a good case for China. First of all, strong public policy must be in place. Unfortunately, California did not do that and enacted a "deregulation law" (1996) that was both difficult to "monitor" but even more significantly gave (sold) energy generation to private companies. By early 2000, these companies realized that they could manipulate the energy market to their advance. Hence an energy crisis is in California with blackouts from 2000 to 2001 and continuing into 2005.

This was not a "perfect storm," as most economists would like to promote. This storm was man-made. Or more pointedly, the energy crisis in California was due to the complete change from public monopolies to private ones. The results are well known. See the first edition of the coauthored book Agile Energy Systems *(Elsevier, 2004)* for more details. But more significantly, the subtitle to the book, which is our focus here: "global lessons from the California energy crisis."

What was learned in California applies to other nation-states. China is an excellent example. Let us consider three major lessons: (1) Agile or flexible energy systems are key to any modern and developing economy. That means that communities such as towns, villages, and cities must have their own power and energy sources. Even within large cities like Shanghai, there needs to be districts and areas with their own independent power. However, these new local or distributed energy systems must be sustainable and built upon core systems that almost exclusively use renewable energy. Below is an example of such systems on residential homes from Japan's Sharp Solar Corporation who call them "zero-emission homes."

A good example of distributed energy generation can be seen by a China company, Sun Chiller, who are marketing and selling their products in California for large buildings and commercial complexes. The economic opportunities are expanding rapidly. In Beijing is Hanergy Group, who produces thin film solar. This is the Hanergy Research Center which is all covered by their thin film solar panels. 83% of their eight buildings in this location are powered by solar. Hanergy has produced three cars that are all-solar powered (http://www.hanergy.com/showCar/carshow.html).

The company now has an office in Silicon Valley in Northern California and more.

The basic lesson is that there needs to be on-site or local power generation as a primary source of energy along with a central grid backup system. Once these "agile systems" are created, then the dependency upon oil and gas or traditional energy sources are reduced as well as renewable energy sources will not pollute the environment and cause atmospheric and climate problems. Another Chinese-based company is SunTech, which produces solar and photovoltaic systems for buildings.

If there are to be growing and new communities that have energy, waste, water, and transportation needs, then they must have integrated and sustainable resources. Otherwise there will be complexes like this one in Riverside County, which is east of Los Angeles in Southern California. Riverside is the fastest growing residential home market in California. This model for development must stop. As one journalist reports, "California looks ahead, and doesn't like what it sees" (Dean Murphy, Associated Press, May 29, 2005).

Agile energy systems: on-site and local power generation

One way to capture this idea is seen in another article in The Economist *(May 11, 2004) which they call the "Energy Internet." That is an agile energy system that combines on-site power generation with a backup central power grid. Some scholars and commentators like the concerns for energy infrastructure to the digital or wireless technological revolution, whereby there is no need for tradition install and hence stranded capital costs but ones that are flexible and "leapfrogged" into the future. Hence for energy, the analogy to wireless communications is to skip the need for landlines. And instead go to a mix of diverse energy supplies through the central grid along with remote or distributed regional energy generation.*

Emerging advanced technologies: leapfrog into circular economics

The second opportunity is energy in terms of fuel for vehicles. One report again from Shanghai notes that parts of China are facing severe fuel shortages, sparking hot tempers and long lines at the pump, reminiscent of 1970s America. This is similar to what has happened in the United States. With skyrocketing fuel costs, people are lined up at petrol stations for cheaper prices. China's fuel consumption and production are outlined like this.

Power Hungry

As China's energy consumption increasingly outstrips production, the government is enacting emergency measures to help avoid acute fuel shortages that threaten to ignite economic and social disorder. At right, drivers waited for hours Monday to fill up at a gas station in Dongguan.

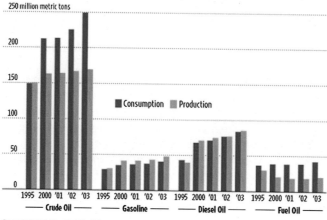

Source: National Bureau of Statistics, China

Source: Wonacott, P., August 17, 2005. Gas lines and growing pains: China's fuel shortages add to pressure to end central planning. Wall Street Journal.

Yet as noted by the New York Times *recently in China, there is a rapidly increased consumption of passenger vehicles. International automakers like this rapidly growing market since it means more sales and higher profits. However, here again there are lessons learned from the California energy crisis.*

Most significantly, energy for commercial, residential, and public buildings is a parallel concern and competition for fuel for vehicles. Furthermore, stationary energy is more polluting today than vehicles' emissions. That is due in part to the California leadership. Public policy was established in the early 1990s creating zero-emission vehicle standards for the State. Separate districts were created to acknowledge the different regions in the State. China is facing similar situations.

As an article about General Motors (GM) indicated sales and competition in China (August 9, 2005) from Liuzhou, China, the partnership with Wuing has placed GM as the top automaker in China.

However, the article and later ones all focus on GM making minivans for USA $5000 is a strategic mistake for China. The larger and less fuel-efficient vehicles mean more demand for fuel to operate. The increased demand means even more dependency in China on fossil fuels that it does not have. And that China should not want to create. Instead, China stands at the cross roads for energy development and production. As noted above, transportation and stationary energy resources (supplies) and usage (demand) are very much intertwined. Herein lies the economic opportunity. China has the opportunity to move ahead into the future now rather than simply copy or mimic what large companies are producing and selling in the United States or EU.

Hybrid technologies are the near term future for China. From The Economist *(December 2, 2004), the basic issue is that the most popular cars in America due to the high gasoline prices are hybrids that combine gasoline and electric power. As noted in this diagram that Prius from Toyota which gets about 45 miles per gallon of gasoline uses "regenerative braking" technology to recharge its batteries from which an electric motor operates the car. This means that the gasoline motor does not need to operate all the time and instead in traffic or city streets or short distances the electric motor operates.*

Along with the "trade" wars the costs for cars will rise as will the fuel for them. Hence all-solar-powered cars and busses as well as trucks and trains will be soon for everyone.

Hybrid technologies: leveraging resources for a "green" economy

The car costs about the same as regular four-door passenger car and performs as equal with most cars. In mid-2005, more and more automakers are producing their own version of the hybrid. While at first the Japanese took the lead, now European automakers are producing these cars for the United States and EU markets. Moreover, the larger SUVs and pickup trucks are not selling so the pressure is on for US automakers to manufacture the hybrids. GM and

Daimler Chrysler announced in August 2005 that they had just formed an alliance, recognizing the market shift.

Finally, new economic opportunities exist in renewable energy sources. The American Council for Renewable Energy (ACORE) held in conjunction with EURO Money in London its Second Annual Renewable Renewable Energy Finance Conference in New York City. The meeting attracted twice the number as the First Conference with a large finance and energy company sector representatives, who focused primarily upon wind power. However, it was clear that solar, geothermal, and other renewables would soon be as significant finance areas. In fact, this sector looks so promising that ACORE and EURO Money will be cosponsoring the same conference in Beijing in November 2005.

One of the key issues is the need to have government policies that set "goals" and standards. California and other states have passed legislation for Renewable Portfolio Standards (RPS) that in most cases double or triple the percentage of renewable energy generation within 5–10 years. Below are some statistics from The Wall Street Journal *on how "energy initiatives gain power in some states" (John Fialka, June 8, 2005:A4). What is critical here is that these RPS laws are all from states and not the federal government who, even with its new Energy Bill (August 2005) did not set any national goals or RPS.*

Nonetheless, financiers and investors see profits from renewable energy. Another article in The Wall Street Journal *noted that "clean energy has investors seeing green" (Gregory Zuckerman, June 23, 2005) best shows the dramatic increase in only just a year and a half. The clear indications are for more finance and economic development to be made available in renewables.*

Renewed Energy

KFX's daily share price on the American Stock Exchange

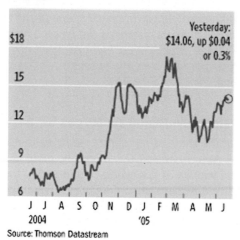

Yesterday:
$14.06, up $0.04
or 0.3%

Source: Thomson Datastream

Again, the focus with investors today is on wind. China has been a direct recipient of those investments from companies like Vestas and NEG Micon, which merged with Vestas in 2004. Both Danish companies have been active in China developing wind power, but not just for central grid transmission as shown below as this one in Middelgrunden Wind Park near Copenhagen, Denmark, but also for smaller communities and even other farms.

The Danish are not alone. Germany and Spain have been world leaders in wind technology development. The United States on the other hand has been slower in developing this technology although it was a world market leader in the 1980s. The only wind turbine manufacturer is GE Wind in California with operations in Europe. Other smaller manufacturers are growing and developing new technologies but still not global. They will be soon, however. As wind power is viewed as a common environmentally friendly source of energy, such as pictured below in Boston Harbor, Massachusetts, the industry will develop as it has done in Europe.

In summary of the main points in this section, it is important to note that the economic concept of hybrid technologies is important if not the most significant. Many recent articles and studies are being published on this point. As the last picture demonstrates with the author, here is a hydrogen fuel cell bus from Daimler near Stuttgart, Germany, in the spring of 2001. This bus was running in the region and now through the public policies of the EU has 33 of them in 10 European cities.

Conclusions

China has the opportunity now through its "social capitalism" public policies to move ahead with standards and protocols that require energy and environmentally sound technologies. Contrary to pressure from the WTO, there is no problem with government involvement in markets. That is exactly what the United States and EU do now. They may not call it public market driven but certainly space travel, world conflicts, and homeland security are all government sector markets.

What is critical for China is to see the environmentally friendly technologies as being a "leapfrog" in economic development. Such a model does not follow the United States and EU economic development pathway, but goes beyond or over it now rather than in stages or a set series of events. A good example is the telecom industry, and especially mobile phones and wireless communications. China and other nations have leapfrogged this industry already.

Similar to mobile phones that allow telecom communications without landlines, China can build energy and environmental infrastructures for the future now. Such infrastructures will require public policies such as "green hydrogen" (that is hydrogen developed from renewable energy sources) in the energy and environmental sectors. In the meanwhile, there will be hybrid vehicles for transportation, which are far more efficient than current automaker productions being advanced in Detroit.

The future for China and globally will be based on environmentally friendly technologies so as not to demand more conventionally available fuel supplies. The nation and regions of China can be more agile and seek energy from renewable sources for a cleaner environment today.

Cities create their benchmarking reports by aligning the key indicators or areas of interest (e.g., energy, water, waste, renewables, transportation, etc.) noted from the baseline assessment. Cities can compare sustainability goals, targets, strategies, standards, and practices across multiple sources to establish their own plan.

Developing an outreach plan to engage stakeholder can provide important legitimization and public buy-in for cities developing a sustainability plan. Stakeholder outreach during the city planning process is important to maintain broad community understanding and support. Actively engaging the communities and stakeholders in discussing aspects of the sustainability plan helps to build a collaborative open dialogue between the city, the public, and key stakeholders interact with one another.

Goals can be developed in coordination with governments and businesses, setting overall goals and targets and a roadmap for creating a sustainability plan. This process could involve meeting with local civic and business organizations, holding public open houses and focus groups, or developing online surveys and meetings. Working with a broad range of stakeholders is important for identifying and engaging community leaders who often become spokespeople, letter writers, and campaign advocates as the process of developing a sustainability plan moves forward.

People need to be involved and interact with one another as individuals and in their family roles of husband, wife, sons, and daughters as well as work, government (at all levels and volunteering to help others). Illustrated below how "interactionism" works:

Additionally, cities and transportation planners need to recognize and accommodate the needs of all public transportation users, including families, older adults, children, and people with disabilities. Inclusion and accessibility for all riders strengthen the system and increase ridership. The goal of multimodal transportation is to offer a variety accessible transport services to meet the multitude of transportation needs.

Developing a sustainability plan does all of this as it should include an evaluation component. Measuring the effectiveness of a sustainability plan is borne out when measurement and evaluation are considered part of a wider iterative process of measuring and reporting and embedding the implementation process into the plan. Above all, the process is not linear. It is circular and even global involving everyone.

Finance, economics, and energy: SMC green development

7

Global green sustainable smart green cities and communities: components for city and community-based green development plans[1]

Woodrow W. Clark II, MA[3], PhD

Founder/Managing Director, Clark Strategic Partners, Beverly Hills, CA, United States

Introduction

Agile energy systems (Clark and Bradshaw, 2004) are becoming the new "norm" and standard for all energy systems around the world. Agile means flexible as there need to be central power grids but also on-site or distributed power systems too. In short, energy production and transmission needs to be flexible or agile: combing both sources of energy.

Over the past decade, Clark has written 11 books and has been working with various national and local governments including the UN B-20 Task Force for its Final Report on September 2–3, 2016 (B-20, 2016) to develop a nearly global plan to ensure there is renewable energy available for the world's future needs that are sustainable and environmentally sound. These efforts are on a grand scale and involve the cooperation of governments and industry together as Clark and Cooke show in their books on The Green Industrial Revolution (GIR) in English 2014 and Mandarin 2015 (Clark and Cooke, 2014, 2016). Although it will take years, even decades to fully implement the GIR, there needs to be discussions and plans for individuals, families, and small businesses, and cities on what they can do today. The book by Clark and Cooke, 2016 titled Smart Green Cities (SGCs) with cases and examples on how cities are becoming "carbon neutral" around the

---

[1] Woodrow W. Clark II, PhD. is a Qualitative Economist and worked with Larry Young on this paper. Clark can be contacted at: wwclark13@gmail.com

Sustainable Mega City Communities. https://doi.org/10.1016/B978-0-12-818793-7.00007-X

world. There needs to be local concern for inefficient energy, control of waste, and means for energy production and consumption. Actions and solutions always start at the local level—from individuals, families, and communities.

The result of the UN G-20 Summit held in Hangzhou, China on September 3—5, 2016 with its Executive Report (G-20, 2016) highlighted the need for both power systems: central gird and on-site distributed energy systems. The results are seen as a "global energy interconnection District Grid" (GEIDG, 2016). Indeed, the term, "interconnection" as well as reference to "green industrial revolution" were key concepts made by Chinese Premier XI in his speech (XI, September 2, 2016), but as green "development" and not "revolution" (XI, 2016) are occurring around the world at not only G20 countries, but many other nations in South America, Africa, Middle-East, and Island Nations.

What this green development "revolution" requires, if there is any hope of success, are the actions of billions of individuals to start taking small local steps, which will accumulate to produce a fundamental change to protect the earth and reverse climate change. Above all, the world is "round" and not flat as with conventional western economics balance between supply and demand with no role for government as they are the "invisible hand." As the data shows below, the world is round and we are all subject to its problems and impacts from climate change to other factors:

Selected Significant Extreme Weather Events, 2014-2015

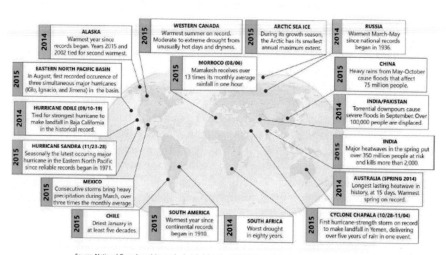

Source: National Oceanic and Atmospheric Administration (NOAA), State of the Climate Reports, 2014, 2015.
Some events were influenced by an unusually large El Niño pattern that emerged in the last half of 2015.
Reproduced from National Oceanic and Atmospheric Administration, State of the climate reports, 2014,2015.

Climate problems have now motivated people and governments around the world to take action—and provide funding. There are now many efforts around the world to build SGCs that will utilize, as seen below, technologies available today such as solar panels, along with a hybrid or electric cars parked in the building garage being refueled or recharged. Additionally, these smart green cities utilize emerging Smart Home devices such as Google's, NEST, Apple's ihome, Amazon's Eco (Alexa), Samsung's X, and Microsoft's X. However, these smart technology examples have one problem: they do not communicate with each other.

Google, for example, started in 2012a "Tech City" program in all 50 American states (Google, 2012). The criteria for a "tech" city is needed but also connected to the city and its infrastructures being connected. The future of smart homes, buildings, and complexes in every city depends upon intercompatibility, such as can be found with the Eco Automation's Hub and cloud-based solution that have universal communications protocol along with top notch energy security (Rice, 2016).

These smart and green devices will control efficient light emitting diode (LED) lighting, HVAC, as well as security and communications systems. The technology exists today to speak to your system and say something like "I'm going to bed" and the system turns off the lights, lowers the A/C, locks the doors, and turns on the alarm. Combine this with a community wide management system to have millisecond demand response. Moreover, you can see how fast this efficiency will accumulate to have a significant positive impact. This is an example of agile energy systems that use readily available energy efficiency technologies combined with central power grids and on-site energy generation, all with storage for emergencies, security, and safety.

Agile Energy System – General and On-Site Power Generation

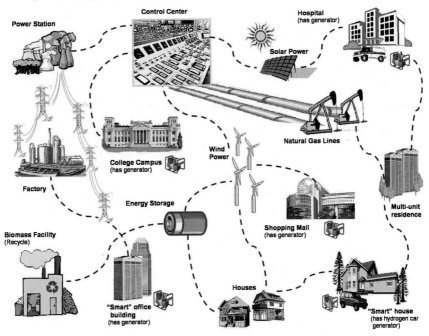

Reproduced from Lund, H. (2014). Analysis: 100 Percent Renewable Energy Systems. In Renewable energy systems: A smart energy systems approach to the choice and modeling of 100% renewable solutions. Elsevier.

Individual home owners and local businesses in a community need to have energy conservation and efficiency under "one roof" combined with on-site power and storage that are the beginnings of smart green cities. The new global energy power systems' model will be "agile" (that is, "flexible") because it can accommodate both green energy on-site power generation and grid-connected power with extremely smart proactive energy conservation technologies. See, for example, the use of hydrogen for on-site power and storage (Kari and Cao, December 2016). These agile systems combine renewable energy, sensors, and wireless Internet connections and similar technologies that direct market mechanisms. This type of energy system is a new economic model that is part of making cities green and smart (Clark, 2017).

Conventional western economics always assume that all efficiency efforts cost too much, or are unproven, or require savvy investigation to avoid buying the wrong technology (Clark and Fast, 2008). The evidence is just the opposite by the fall of 2016 when economists acknowledged that LED bulbs had become "disruptive." LED lighting is also financial efficiency and has matured, such that prices have come way down, performance is much better, and there are third party agencies that can provide the buyer a level of confidence in what they buy. That is not to say it is a "no-brainer." Where anyone can simply go to Home Depot or Lowes for a great deal, it still makes sense for the small to large business owner to seek the advice of energy consultants such as energy efficiency experts (CITATION). E3 has experts who can sift through the dozens, if not hundreds, of choices and specifying and sourcing lamps can be a daunting task. The industry now suffers an abundance of riches … and it takes an experienced energy team to know what is the proven latest and greatest, which might change on a monthly basis.

One nontech financial efficiency is deregulated power contracts. Most states should have some form of deregulated power by the year 2020 and this means lower power costs for grid power, and increased opportunity to buy green/clean power from the grid. All energy demand starts at the local level with on-site power for homes and businesses AND access to grid power. The Power Company is another example.

Regulators are now implementing carbon dioxide regulations and even taxes (as was done with tobacco) to stop pollution. Meeting the challenges of supplying energy for increasing demand, while reducing carbon emissions, calls for more complex and creative solutions based on local and national plans. The change starts with efforts to increase energy efficiency, use renewable energy generation, and create infrastructures for water, waste, and transportation that are integrated. Together these changes will help the way people live and plan for all kinds of activities from using electricity to getting to work. Case in point is illustrated later with the use of LED lights and how they impact energy use in buildings.

For the last decade, the growth in the LED industry has been incredible. For both the consumer (lower costs for LED bulbs) and the manufactures, the costs have come down dramatically. Today as the chart below from Goldman Sachs Global Investment shows (August 2016), LEDs out-perform all the other smart green industrials in terms of market share and growth. This does not dismiss the solar and wind industries, but does make it clear that energy conservation and efficiency is critical to all areas of reducing carbon emissions and produce immediate financial savings.

LEDs are just one example of what a home or business can do right now. In every aspect of your home or facility, there are improvements that can be made today, the materials for your roofs, walls, and windows, air conditioning controls, sensors, and monitors, all coming together for the smart home, smart building, and smart city (www.Nuralis.com). One US company, Premier Holding (PRHL), is pioneering this effort with a focus on individual homes, businesses, and communities.

In a recent article about the "green industrial revolution" in Southeast Asia (SEA), Clark noted both the need for this green development as well as its impact. SEA is primarily made up of over 18 developing countries. For the last 2 decades, Japan has been supporting and helping the SEA nations via APO and other opportunities for the countries, cities, and businesses (APO, 2016). China now is looking into the region for both economic trade growth and establishing cooperative relationships. One in particular is Singapore, which is an "Eco-City" and has now partnered with several cities in China to create ecocities throughout China (Chung, 2016).

Governments on all levels and utilities (central grid and on-site power) are all promoting energy conservation and efficiency that involve new appliance standards and energy codes or incentive and educational programs. Typical programs are as follows: rebates for the purchase of energy-efficient products, energy auditing services, and incentives for upgrading the operational efficiency of buildings, processing plants, and other facilities. There are also programs that encourage the purchase of new high-performance homes, and others that provide design and engineering support for qualifying business and home buildings.

More problems come with only a central electricity grid unless they are transformed into a smart, responsive, and self-healing digital network—in short, an "energy internet" noted earlier is needed. Moreover, the agile energy grid (both central and on-site energy systems) needs to be implemented sooner than later. The US Department of Energy put it very well in the Revolution Now Report from the summer of 2016:

Revolution … Now—2016 Update.

Decades of investments by the federal government and industry in five key clean energy technologies are making an impact today. The cost of land-based wind power, utility and distributed photovoltaic (PV) solar power, LEDs, and electric vehicles (EVs) has fallen by 41% to as high as 94% since 2008. These cost reductions have enabled widespread adoption of these technologies with deployment increasing across the board.

Combined, wind, utility scale and distributed PV accounted for over 66% of all new capacity installed in the nation in 2015. Total installations of LED bulbs have more than doubled from last year, and cumulative EV sales are about to pass the half-million mark. See them and other environmental technologies below:

These technologies are now readily available, and countries have already begun to reap the benefits through their increased adoption. As these clean technologies are broadly deployed, there is a reduction in the emissions that contribute to climate change, the air we breathe is better quality because of a decline in air pollutants, and we are expanding economic opportunities for American workers and manufacturers. In 2014, the manufacturing sectors for wind turbines, photovoltaic panels, lithium ion batteries, and LEDs have added $3.8 billion dollars in value to the US economy.

As we continue to advance international action on climate change under the Paris Agreement—which established a long-term worldwide framework to reduce global greenhouse gas emissions—these five technologies have and will play a critical role in providing opportunities to reach global climate goals.

Conclusion

Renewable energy generation, along with cutting edge efficiency technologies, is the foundation for a sustainable community, and the heart of the Green Industrial Revolution.

Although most countries have or are developing large-scale solutions to our critical problem of energy generation and conservation they will take years to implement. Although all 200 countries in the world agreed in October 2016 to the UN G-20 Summit Conclusions, most nations will take a long time to implement what they must do now. Only a few billion people can do today with existing technology. Indeed, these large-scale plans require the cooperation and proactive efforts of individuals and communities. There is no reason to wait and see what these plans do in the coming years, with a few phone calls you can begin to make a difference, and together, the world can and must change.

References

Chung, R.K., September — October 2016. Quality, not quantity of growth can lead to green goals: low-carbon economies will become a realty only by marrying long-term gains with short-term economic interest. APO News Journal 6—7. http://www.apo-tokyo.org/publications/apo_news/september-october-2016/.

Clark, Woodrow W., Bradshaw, Ted, 2004. Agile Energy Systems. Elsevier Press.

Clark, Woodrow W., Cooke, Grant, 2015. Green Industrial Revolution. Elsevier Press.

Clark II, Woodrow W., Cooke, Grant, March 2016. Smart Green Cities. Routledge Press.

Clark II, Woodrow W., 2017. In: Sustainable Communities Design Handbook, 2nd Ed. Elsevier Press, Fall.

Clark II, Woodrow W., Fast, Michael, 2008. Qualitative Economics: Toward a science of economics. Coxmoor Press.

Analysis and Policy Observation (APO) Melbourne, Australia at http:mailto:webmaster@apo.org.au" webmaster@apo.org.au.

E3 References

Google, "Tech Cities" Started in 2012. 2012. at: https://www.google.com/economicimpact/ecities.html.

Rice, S., September 21, 2016. Web. 01 October 2016. Integrating climate change into national security planning. The White House.

XI, J., September 2—3, 2016. Keynote Talk to the B-20 Task Force Summit, Hangzhou, China. Notes from translation of 1 hour speech.

Cities hold the keys to greener, more efficient homes: decisive policies are critical for achieving climate goals in residential buildings[2,3].

Michael Gartman

Professor, Poli-Technical University Milan, Italy

Our homes may be a source of safety, comfort, and stability—but they also represent a considerable slice of our country's carbon emissions (19%, according to the latest estimate from the US Energy Information Administration).[4] Addressing this piece of our energy system is essential to achieving our climate goals. Perhaps more importantly, improving the efficiency of our homes is a powerful tool for addressing the energy burden that disproportionately impacts our low- and moderate-income (LMI) communities.[5] Fortunately, cities around the United States are collaborating and taking aggressive action.

Policy challenges

Several hurdles stand in the way of decisive policies addressing residential energy efficiency:

- A residential market mostly based on individual owners makes scaling change particularly challenging
- Local context (including climate, architectural style, and constituent priorities) is especially influential for residential buildings, elevating the importance of a customized approach
- Smaller buildings (with smaller budgets) can make it more difficult to achieve cost-effective project results with a customized approach
- Poorly designed policies risk exacerbating housing affordability issues or slowing economic growth

Yet the public is increasingly demanding action from our governing bodies. Polling performed by Stanford University researchers recently found that 68% of

---

[2] © 2019 Rocky Mountain Institute. Published with permission. Originally posted on RMI Outlet.
[3] To learn more about the outcomes of RMI's past working groups or to collaborate with other cities to implement home energy labeling programs, or if you have ideas for the focus of future sessions, please reach out to Jacob Corvidae at jcorvidae@rmi.org
[4] https://www.eia.gov/tools/faqs/faq.php?id=75&t=11.
[5] https://aceee.org/research-report/u1602.

Americans believe that our government should do more to address climate issues.[6] A study from the Demand Institute identified energy efficiency as the number one unmet need in residential housing, beating out other categories such as updated kitchens and finishes, privacy, and even safety.[7]

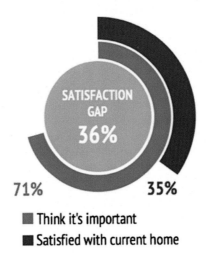

SATISFACTION GAP
36%

71% 35%

■ Think it's important
■ Satisfied with current home

The satisfaction gap for increased efficiency

Reproduced with permission from www.rmi.org.

Cities taking the lead

More and more cities across the country are embracing climate and sustainability goals, and are looking for solutions to build a more efficient economy, more affordable and equitable housing stock, and a more secure energy future. These cities are increasingly relying on collaborative efforts such as the American Cities Climate Challenge, Net Zero Energy Coalition, or City Energy Project to address the daunting challenge of achieving aggressive goals in residential housing and other hard-to-reach market segments. These collaborative opportunities ensure that cities can share lessons learned, navigate common policy pitfalls, and build on each other's successes.

Leading cities across the country are already providing examples of what is possible. Santa Monica, California, implemented a zero-energy performance code for single-

---

[6] https://earth.stanford.edu/news/public-support-climate-policy-remains-strong#gs.4x4gft.

[7] http://demandinstitute.org/demandwp/wp-content/uploads/2014/12/top-10-housing-desires.pdf.

family homes in 2017, over a year before the California Public Utilities Commission's decision to enforce this level of performance statewide.[8] The City of Boulder adopted the country's first efficiency standard for rental housing, inspiring others to follow suit.[9] In February, Minneapolis passed a set of three major energy disclosure policies addressing multifamily buildings, rentals, and newly listed homes.[10]

These cities and more are driving toward aggressive climate goals with targeted action. How can we ensure these successes continue to be replicated and scaled?

Tackling energy issues with collaborative cohorts

At Rocky Mountain Institute (RMI), we are working to amplify these pioneering city initiatives by bringing together collaborative cohorts of city policymakers focused on high-priority energy issues. In 2018, we offered two such working groups that provided city policymakers with the tools, resources, and collaborative problem-solving environment necessary to hone and enact the policies highlighted in two recent RMI reports[11]:

- **Cohort #1: Policies for Zero-Energy (Ready) New Homes** utilized the analysis and insights provided in our *Economics of Zero Energy Homes* report to inform aggressive new construction policies. These policies stand to optimize energy use at the beginning of a building's lifecycle, minimizing cost and maximizing long-term impact, while reducing the total cost of homeownership and bolstering a new labor market—both potential boons to LMI communities.

---

[8] https://www.santamonica.gov/press/2016/10/27/santa-monica-city-council-votes-in-the-world-s-first-zero-net-energy-building-requirement-implementation-begins-in-2017.
[9] https://bouldercolorado.gov/plan-develop/smartregs.
[10] http://news.minneapolismn.gov/2019/02/15/minneapolis-require-residential-energy-disclosure/.
[11] Download the reports here: https://www.rmi.org/insight/economics-of-zero-energy-homes/ and https://www.rmi.org/how-cities-can-ensure-better-rentals-for-everyone/.

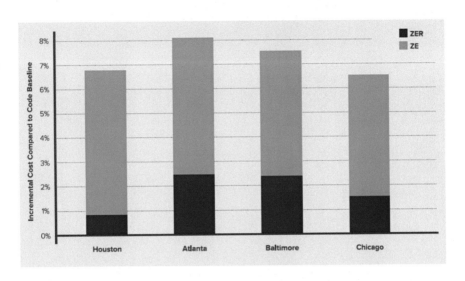

RMI's modeled incremental costs for zero energy and zero energy ready homes

Reproduced with permission from www.rmi.org.

- **Cohort #2: Minimum Efficiency Standards for Rentals** helped cities work toward implementing the rental licensing policy outlined in our *Better Rentals, Better City* report. These policies are particularly important for LMI communities because they stand to enforce a minimum level of performance and comfort in homes that have often been neglected due to the tenant−owner split incentive.[12]

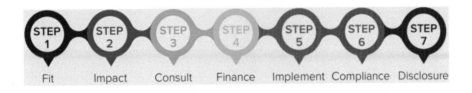

RMI's seven steps to developing minimum efficiency standards for rentals

Reproduced with permission from www.rmi.org.

---

[12] https://www.greentechmedia.com/articles/read/a-graphic-that-illustrates-the-problem-with-split-incentives#gs.59msw2.

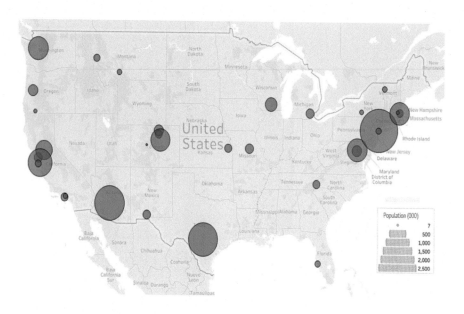

Cities served by RMI's 2018 cohorts

Reproduced with permission from Petersen, Alisa, Michael Gartman, and Jacob Corvidae. The Economics of Zero-Energy Homes: Single-Family Insights. Rocky Mountain Institute, 2019. www.rmi.org/economics-of-zer-oenergy homes.

The cohorts served 33 cities representing over 18 million constituents across the United States. These cities are now using each other's experiences and ideas to inform aggressive action, including:

- Washington, D.C., is currently working toward adopting one of the country's most comprehensive residential zero-energy policies, having recently secured funding for pilot project incentives and convened over 100 local stakeholders to guide program design.
- Ithaca, New York, plans to bring legislation for zero-energy new homes to City Council later this year.[13]
- Policymakers in Bozeman, Montana, are using the cold climate addendum to RMI's *Economics of Zero-Energy Homes* report (which was produced based on demand from our cohort members) to validate the viability of stretch codes and other net-zero-energy-ready policies.[14]

---

[13] http://www.ithacagreenbuilding.com/.
[14] Learn more about the cold climate addendum here: https://www.rmi.org/how-to-cost-effectively-withstand-the-next-polar-vortex/.

- Ann Arbor, Michigan, and other cities are leading the charge to launch new rental efficiency standards.
- Somerville, Massachusetts, is exploring a rental-licensing program to enable efficiency programs.[15]

These cities are building upon leading edge examples and stand as precursors to a countrywide shift toward adopting two sets of policies that can ensure that homes everywhere are more efficient, more comfortable, and more affordable for residents.

This year, in partnership with Earth Advantage, RMI is offering cities free technical support to help them develop home energy labeling and disclosure policies, which cities are increasingly looking toward implementing to provide the actionable information necessary to drive consumer action and unlock economic development in the residential sector.

This spring (2019), RMI will do an initial "deep dive" into home energy labeling policies that will highlight the leading examples of cities such as Portland, Oregon; Austin, Texas; Berkeley, California; and Minneapolis, Minnesota, and synthesize the best practices and necessary steps for implementation. After using that information to mobilize resources and stakeholders in their city, participants will collaborate with each other through a 6-month series of problem-solving workshops that harness the experience of expert practitioners and incorporate a toolkit of resources built specifically to support policy implementation.

Cities across the country are stepping in with impressive leadership in the face of the federal government's withdrawal from the Paris Climate Agreement. It is largely through their leadership that we can continue to move toward our climate goals while ensuring safe and healthy homes for all.

---

[15] https://www.somervillema.gov/sites/default/files/somerville-climateforward-plan.pdf#page=40.

City of Santa Monica, California

The plan includes major investments and simple things everyone can do in their day-to-day lives to make a difference. It advances existing initiatives to enhance community wellbeing, smart city innovation, transportation, public health, and social equity.

The Climate Action section focuses on eight objectives in three sectors:

- Zero Net Carbon Buildings
- Zero Waste
- Sustainable Mobility

The Climate Adaptation section focuses on community resilience to climate change through four sectors: (1) climate ready community, (2) water self-sufficiency, (3) coastal flooding preparedness, and (4) low carbon food and ecosystems.

Council directed staff to modify the plan to include additional measures and information. A final draft will be released shortly. See the current draft plan here.

The Santa Monica Community Recycling Center is closing in June 2019. The closing will not impact curbside or commercial collection services. Drop-off recycling is available in alleys throughout Santa Monica in the network of the large blue community shared recycling containers. Options for the return of the Santa Monica buy back center are currently being evaluated. Visit www.smgov.net/r3 for updates.

Applicants need to demonstrate that they have the ability to properly install the sensor in an optimal location and are willing to participate in the program. Applications must be submitted by June 30. Applicants will be selected by July 15. The Office of Sustainability and the Environment will select 10 applicants with a goal of having an equitable distribution of sensor locations throughout the Pico and Sunset Park neighborhoods. To learn about the research project by The Pardee RAND Graduate School, click here. To apply, click here.

This workshop is primarily for architects, energy modelers, designers, builders, and developers, but all are welcome. At the workshop, City staff will present the final draft Reach Code language that maintains Santa Monica's green building leadership and are deemed cost-effective from recent studies. Join us on **June 11th 10:30 AM–12:00 PM** to learn about the City's process developing a Reach Code and what concept will be brought to City Council and the California Energy Commission for approval.

The City of Santa Monica's Office of Sustainability and the Environment along with Sustainable Works congratulates the winners and participants of the 11th Annual Sustainable Santa Monica Student Poster Contest. This year's theme, My Sustainable School, educated students on the recently adopted Sustainability Plan for SMMUSD, which is organized into eight categories: climate, education and engagement, energy, water, waste, transportation, food and nutrition, and operations.

Solar Santa Monica is a free service for residents and businesses looking to go solar. Solar experts are available to provide you unbiased technical advice to help you navigate the changing rules, incentives and financing options. If you have interacted with the Solar Santa Monica program, we would like to assess the program's helpfulness to you. Take our survey and tell us about your Solar Santa Monica experience.

Did you know that people are more likely to invest in solar power when they know their neighbors have solar? Did you know that Rooftop Solar has been identified as one of the best solutions to climate change? If you are a fan of solar power, Climate Action Santa Monica wants to hear from you! This local volunteer group is helping Santa Monicans learn how solar can lower utility bills, keep your family going even when the grid is down, and improve the value of your home.

Please email climateactionsantamonica@gmail.com to learn more about volunteer opportunities and becoming a Santa Monica solar champion!

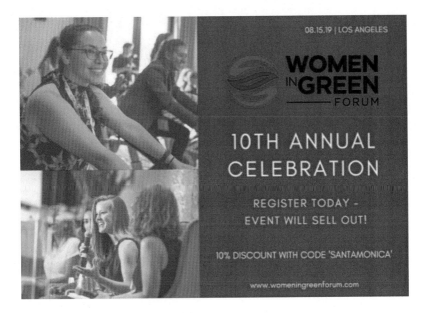

Women climate action

Further reading

Alanne, K., Cao, S., December 2016. Zero — Energy Hydrogen Economy (ZEH2E) for Building and Communities Including Personal Mobility", Dept of Energy Technology, Alto University, Finland, Renewable and Sustainability Energy Reviews. Elsevier Press. https://doi.org/10.1016/j.rser.2016.12.098.

APO News Journal 46 (5), September—October 2016. http://www.apo-tokyo.org/publications/apo_news/september-october-2016/.

B-20 Task Force Final Meeting, Hangzhou, China, September 2—3. http://en.b20-china.org/.

Clark, W. W. II, T. Bradshaw, 2004. Updated for Elsevier Press, 2017. http://www.amazon.com/s/ref=nb_sb_noss?url=search-alias%3Daps&field-keywords=Agile+Energy+Systems.

Clark II, W.W., 2012. Lead author and Editor. The Next Economics: Global Cases in Energy, Environment, and Climate Change. Springer Press.

Clark II, W.W., Cooke, G., December 2014. The Green Industrial Revolution. Elsevier Press (in English). http://www.amazon.com/The-Green-Industrial-.

Clark II, W.W., Cooke, G., 2016. Smart Green Cities. Routledge Press. https://www.amazon.com/Smart-Green-Cities-Toward-Neutral/dp/1472455541/ref=sr_1_1?ie=UTF8&qid=1 472500856&sr=8-1&keywords=Smart+Green+Cities+%28Clark+and+Cooke%29+ 2016.

Clark II, W.W., Fast, M., 2008. Qualitative Economics: Toward a Science of Economics. Coxmoor Press. Updated and Published in 2017.

Clark II, W.W., Cooke, G., Jin, A.J., Lin, C.-F., August 2015. Green Industrial Revolution in China (Mandarin). Ashgate and China Electric Power Press. http://www.sgcc.com.cn/ ywlm/gsgk-e/gsgk-e/gsgk-e1.shtml.

dangdang. http://search.dangdang.com/?key=%C2%CC%C9%AB%B9%A4%D2%B5% B8%EF%C3%FC.

G-20 Summit Executive Summary, Hangzhou, China, September 3−5, 2016. http:// g20executivetalkseries.com.

GEIDG, 2016. Global Energy Interconnection Develop Group/Company. Established. www. geidco.org.

Google Tech City Award Winner for 2015 in California is City of Beverly Hills. http://www. canyon-news.com/beverly-hills-receives-2015-google-ecity-award/46603.

Liu, Z., 2015. Global Energy Interconnection. Elsevier Press, Amsterdam.

PRHL www.prhlcorp.com.

University of California, Berkeley. School of Law: Energy and Environment, 2016. https:// www.law.berkeley.edu/research/clee/research/climate/climatc-changc-and-busines.

Circular economy: the next economics

8

Woodrow W. Clark II, MA[3], PhD[1], Danilo Bonato, MBA[2]

[1]*Founder/Managing Director, Clark Strategic Partners, Beverly Hills, CA, United States;* [2]*General Manager, ReMedia, Via Messina 38, Milano, Italy*

Overview

For more than a century now, economics has been following the economic paradigm set by Adam Smith where there is a linear "market supply" and due to "demand." This is the flat linear economic model. Then, government is an invisible hand that is above the flat economic line. All of this has been in theory and NEVER in fact been true. More details are given later as economics must also be "qualitative" so that it defines, explains, and covers the meaning of statistics, graphs, charts, and more. Economics is NOT just a lot of numbers that are "used" or gotten from "sources" that have little value, definition, and explanation what and from where they came (Clark and Fast, 2008 and 2019).

Case in point is the points made below for the definition of "green" cars. As history clearly shows, green (use of solar and electric power) cars are NEW today. Before that for over 3 decades, "clean" energy (which is what the natural gas industry and even diesel-powered cars were called until recently) cars were marketed and sold as good for the environment and our earth. That is and was false information. This is one of the key reasons that some younger USA Congress and Senate members are calling their "Agenda" the "Green Deal" (Ocasio-Cortez and Ed Markey, 2019). They get it. Dramatic changes much like The Green Industrial Revolution (Clark and Cooke, 2009, 2014, 2015 and 2016) that defines what the world must do (starting at home and in your own country) to reverse the changes in the global climate are required.

For example, when was there even a demand for cars, let alone gasoline and now electric cars? None. The European and then USA industries, businesses, and consumers are now producing all-electric cars due to what Henry Ford did at the turn of the 20th Century the Ford Motor Company with assembly mass production of cars to bring down their costs. The car industry started after Thomas Edison had his energy and lighting company started. All of these companies were part of the

First Industrial Revolution. Now today, we call it The Circular Economy—as seen here:

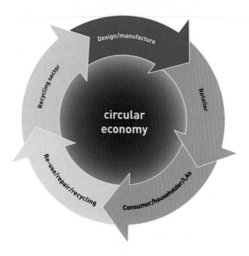

Courtesy of The Waste and Resources Action Programme http://www.wrap.org.uk.

The world is round and not flat. Work on Circular Economics has started due to a report by the Ellen MacArthur Foundation and other partners that will be noted later. A circular economy seeks to rebuild capital, whether this is financial, manufactured, human, social, or natural. This ensures enhanced flows of goods and services.

Introduction

Ford Motor cars until 1923 used fuel for their cars from the family farm land crops that produced juices owned near Detroit, MI. Why did that juice fuel stop in 1923? This is because the oil industry that was growing pressured the USA Congress to outlaw juices from crops for fuel. The oil industry won. Moreover, they continued to get their oil for producing fuel into gasoline for cars.

Today, we have the Tesla all-electric cars. Was there any demand for Tesla cars? No. The demand was not there but has now come so that ALL car makers are producing all-electric cars. Electric cars (Green Car, 2018) are now in 2019 the most popular car. Most of the "million electric cars are sold in the U.S." and over half of those are in California. As Scientific American noted, "Mathematics from quantum mechanics explains how resources in the atmosphere can amplify the bends, making harsh weather even worse" (Mann, 2019) due to the changes in the jet stream and more coming. Climate change is real.

The basic issue is that when electric cars (EV for electric vehicles) were first introduced in the 1990s by the major car companies around the world, they gained a lot of attention. Primarily, these cars were rented or leased. The EV was not popular with the companies that supplied fuel for the gas used in most vehicles (cars, trucks, vans, etc.) so that the EV car companies were "pushed" to stop producing them. In fact, the American cars were "gathered" into large areas where they were destroyed. The EV was replaced with cars that then used natural gas for fuel and after the turn of the 20th century, hydrogen fuel cell cars came slowly into the transportation market.

Electric cars today: then all solar-powered cars is the next economics

The car manufacturers today do not want to make the same mistakes that they made at the end of the 20th century. Hence, the ALL electric car was created with the Tesla Car becoming the leading manufacturer and distributer. Was there a "market" for the Tesla? NO. Was there even a demand for electric cars? NO. In fact, the Tesla name is from the famous scientist at the turn of the 19th century who invented wireless energy (AC) systems. Tesla was in competition with Thomas Edison at that time who wanted DC produced from energy that was produced from fossil fuels such as coal and oil and then distributed in pipes, wires, etc. (Seifer, 1996 updated 2006 and Clark, 2004 and updated 2017).

What happened was that Elon Musk (a successful entrepreneur in other businesses) saw that there was a need for the ALL electric car now and was able to raise funds to manufacture and then promote the sales of his Tesla Car.

Now, General Motors (GM) declared in early December 2018 that it was shutting down five car manufacturing plants with over 15,000 people losing their jobs. The reason was GM decided that it now wanted to manufacture and sell only electric cars, as Green Car (2018):

A new Chinese electric carmaker plans to sell electric cars in the United States via the owner of the defunct automaker Coda. A new Tesla Model S unexpectedly caught fire—twice—after getting a flat tire in California. Moreover, Volkswagen teased a video and images of its new I.D. electric hatchbacks testing in South Africa. All these and more are on Green Car Reports. *Electric car fans have long wondered why automakers do not do a better job of promoting their benefits. Now that automakers are being required to sell more electric models outside of California, Nissan is ramping up those efforts for its Leaf electric car.*

Mullen Motor Company, the California start-up that bills itself as "the affordable electric car company," plans to start selling an electric sports car from China alongside its small neighborhood electric vehicles and (previously) leftover Coda sedans. Moreover, Volkswagen, perhaps growing tired of listening to online skeptics who doubt the company, is serious about selling electric cars.

The Nissan Leaf with a roughly 60-kilowatt-hour battery was expected to make its public debut at the LA Auto Show earlier this month. It did not. That car is expected to appear "very soon," perhaps at the 2019 Consumer Electronics Show to be held in Las Vegas next month.

All of the news and "demand" for electric cars was never linear with supply. The lines of people who went in 2016–17 to put a down payment on Tesla cars never saw in the demand for electric cars in any data supplied about cars, especially all-electric cars. The same was for hybrid cars when they were first (Prius) released by Toyota in the late Spring of 2003. What happened was that the public demand rose and expanded after the Prius. Then later Tesla, which greatly reduced the costs for its cars, has spread its sales around the world (especially in China) where the Tesla is now being manufactured in China as well.

However, the largest renewable energy company in China (Hanergy) located in Beijing went ever further in Circular Economics as it saw the use of thin film solar manufactured for buildings is now being applied to cars. Why? As noted here and then more details so that the sun comes down on cars with thin film solar on the cars that are recharged as they are driven, parked, or even going to and from home and office events (Clark, July 2016)

Although the sun does not shine at night or even during the day due to clouds and rain, the cars get power from the sun and stored in the car. In short, a "solar car" is Circular Economics in action.

Following this, solar car from Hanergy Energy Group is a dramatic technology change for cars has started and the cars will be out soon from a company in Germany. Meanwhile, the China Daily reported at the end of 2018 about the overall car industry in China and USA:

> As California-based Tesla Motors Inc. will start partial production in Shanghai in the second half of next year, and Boeing delivered first airplane from its Zhoushan facility in China earlier this month, business leaders believe that stable China–US relations will help more companies from the United States to maintain robust growth in China. China has become an integral part of the growth plans of many US businesses, not just because of its sheer size but also due to its strategic importance, according to Stephen Shafer, president of 3M China, the Minnesota-headquartered conglomerate. "China is actually becoming the leader in many markets and technologies that we are interested in," said Shafer, referring to digital platforms in China that help companies engage with customers. (Daily, 2018)

Why economics needs to be circular

Case in point is the EU as when (2014) Europe was the year of the circular economy, as the new global paradigm for designing patterns of development of all societies and industries from now to 2050. However, it was not. Europe had the opportunity to again lead the world as it has done in the Green Industrial Revolution along with Japan, Germany, and China providing strategies (Clark and Cooke, 2014 with Anjun Jerry JIN and Ching-Fuh LIN in Mandarin and 2015, English).

President XI of China has adopted these policies, strategies for economics due to his leadership that created and activated what he calls "green development" and "silk road" (XI, 2016). Still, the United States is not pursing a Circular Economy at levels of public policy economics. As Circular Economics is implemented in the circular model paradigm, which is not the linear one of conventional economics with supply or demand. Therefore, the Ellen MacArthur Foundation plans to team up with groups in the United States as it did in China July 2018 ( ). The first Circular Economy (CE) Conference was titled Circularity held in Minnesota, USA (mid-June 2019) and sold out due to the Agenda that featured economic-based groups, startup companies, major high-tech companies whose focus was sharing information; connecting to funding, finance and; resulting in measurable standards for stopping climate change which will dramatically change the conventional USA economy.

China has focused on green development that includes energy, water, air, transportation, and buildings so that their environment is not polluted and the economy (Energy News, January 2019) will flush despite problems with trade and the United States. Case in point is electric cars now with the all-solar-powered cars by Hangery coming in a few years. China already has electric buses (China, 2018). According to the Climate Change Report, China in "A New World" is the "absolute winner" of clean energy transition (January 2019).

China took a BIG step on July 16, 2018 when a meeting Beijing with the EU Circular Economic leaders signed a Memorandum of Understanding (MOU) between the two economic powers:

MEMORANDUM OF UNDERSTANDING ON CIRCULAR ECONOMY BETWEEN THE EUROPEAN COMMISSION AND THE NATIONAL DEVELOPMENT AND REFORM COMMISSION OF THE PEOPLE'S REPUBLIC OF CHINA

In short, the signing of the MOU agreement by the two world "largest economics" would enhance, broaden, and accelerate Circular Economics around the world, which would create a global scale plan and actions to have a paradigm shift to low carbon and a regenerative economics. The Ellen MacArthur Foundation originally funded the Circular Economic research, plans, and implementation, which started with the EU in 2015. The Circular Economics in cities throughout China would make goods and services more affordable for everyone as well as reducing the economic impacts on people, communities, governments, and businesses with lowering carbon emissions, recycling waste, and reversing climate change.

In December 2018, the Ellen MacArthur Foundation released a Report on China for Circular Economics in China's cities that would "make goods and services more affordable for citizens, and reduce the impacts normally associated with middle-class lifestyles." Below is what happens in theory but now more and more Circular Economic results are becoming the new "economic standard" with success on all levels and communities as shown below.

For Europe and the whole world are facing several pressing issues concerning climate change, the environment, society and the economy, which are crucial to the quality of life of our children and grandchildren. Circular Economics does not appear on the demand side of the linear economic paradigm from Adam Smith, but does now as will be described later due to the need for nations, states, cities, and companies all need to see the financial value in economics being circular.

For this reason, in 2013, the European Commission launched a new and comprehensive research program that was innovative called Horizon 2020, which was to be an introduction to focus on concrete solutions to the environmental challenges that create and continue to change the global climate. Today in research from the EU and next in China will show how economics is really round.

Scholars, policy makers, and economics all agree that Circular Economics is of importance that this program is not only convincing from the point of view of research, but that it is also relevant in contribution to the achievement of the

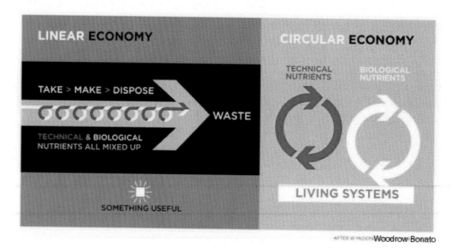

Conventional neoclassical economics circular green economics.

objectives of the European Union, including prosperity, quality of life, sustainability, growth, and employment. The comparison of the two economic paradigms is as follows:

One of the main objectives of Horizon 2020 is to position Europe as leader in the development of a "circular" and "green" economy, based on the concept of sustainability: at this particular moment in time the EU must strengthen its international competitiveness through the productivity of resources and the improvement of the capacity to provide the world with low environmental impact technologies and services, oriented toward an efficient use of resources.

What is critical to understand and apply from this paper is how economics needs to become a science. To do that, economics needs to be both quantitative (which it is primarily today and for over the last 3–4 decades) and qualitative.

Circular economics

Circular Economics is therefore an epochal change that must move away from the conventional neoclassical economy of Adam Smith (Clark and Fast, Qualitative economics, 2008 and second Edition, 2018) to include other aspects in our development models as well as the "market" Above all, Circular Economics is scientific as economic needs to be both "quantitative" and "qualitative" (Q^2E). The qualitative paradigm for economics follows the linguistics paradigm of Professor Noam Chomsky (MIT) (https://chomsky.info/) where he made linguistics that was qualitative also into being quantitative with charts, graphs, data, and statistics. There is more on Q^2E below on how both areas are the key to science.

One of the key priorities for Europe in the circular economy is that of resource efficiency and waste reduction (Zero Waste Strategy). In this field, several *stakeholders* are active. Consider Remedia Company in Milan, Italy (https://www.consorzioremedia.it/en/) as the collective business system of manufacturers for electrical and electronic equipment. Remedia developed a long-term vision to fully exploit the potential of secondary raw materials and also increase the innovation capacity of the recycling sector in the EU to be present in all regions and countries. The point is that products can be made by recycled waste from others and then become recycled once they are dated, need to be updated, and replaced by other newer products.

The business dynamics is linked to the circular economy and thus transformed into a solid pillar of the business growth strategy of the entire European Union and nations around the world. Above all, circular economics creates jobs, businesses, and future investments today while providing education, research, and strategies. The world of industry, in collaboration with governments, must work to address social environmental challenges and generate concrete benefits for individuals, businesses, and their communities.

The Circular Economics system diagram illustrates the continuous flow of technical and biological materials through the "value circle." The application of Circular Economics is now global as on July 16, 2018; China and the EU signed a MOU about collaborating, coordinating, and enhancing a Circular Economic future TODAY.

The circular economy

In a circular economy, the "field" of economics activity builds and rebuilds overall system health, which will soon turn economics into a science. The concept recognizes the importance of the economy needing to work effectively at all scales—for large and small businesses, for organizations and individuals, globally and locally. Circular economics is a key factor in how economics can become a science. Chemistry, physics, engineering, and wifi program are all circular as well, as the sciences need and do connect with one another for many reasons: research, public policy, costs and finances, etc.

Transitioning to a circular economy does not only amount to adjustments aimed at reducing the negative impacts of the linear economy, but also it represents a systemic shift that builds long-term resilience, generates business and economic opportunities, and provides environmental and societal benefits.

Technical and biological cycles

The model distinguishes between technologies and science cycles. Consumption happens only in biological cycles, where food and biologically based materials (such as cotton or wood) are designed to feed back into the system through processes like composting and anaerobic digestion. These cycles regenerate living systems, such as soil, which provide renewable resources for the economy. Technical cycles recover and restore products, components, and materials through strategies like reuse, repair, remanufacture, or (in the last resort) recycling. When areas are related and interactive together, that is how circular economics works.

As shown earlier, the use of land for exercise, relaxation, and fun can also have be good for bike riding and solar-powered cars along the green energy from solar and wind power. The key are the energy needs for the buildings in the area as they can all be served by "on-sit" power that is "distributed" and used locally rather than piped or sent from a central grid. This kind of collaborative and circular economic systems is called **Agile** (which means flexible) **Energy Systems** that was the topic of Clark's first book in 2004 (Elsevier Press), which focused on California energy problems and their solutions.

Then, in 2017, Clark's second edition of the **Agile Energy Systems** book from Elsevier Press was written and published, which focused on energy international global systems in nations, regions, communities, and cities (https://www.elsevier.com/books/agile-energy-systems/clark/978-0-08-101760-9A). Agile energy systems are good for people, businesses, government, and every energy system, as they need to be on-site distributed and central grid together as they serve the entire community, city, and region.

Origins of the circular economy

The theory and practice of circularity has deep historical and philosophical origins. The idea of feedback, of cycles in real-world systems, is ancient and has echoes in various schools of philosophy. It enjoyed a revival in industrialized countries after World War II when the advent of computer-based studies of nonlinear systems unambiguously revealed the complex, interrelated, and therefore unpredictable nature of the world we live in—more akin to a metabolism than a machine. With current advances, digital technology has the power to support the transition to a circular economy by radically increasing virtualization, dematerialization, transparency, and feedback-driven intelligence.

The new economic paradigm: circular economics

There are plenty of opportunities to rethink, redesign, and extend the way the everyday workplace actions ranging from creating and then industrializing cars to computers to fruits, vegetables, and the products that people produce and use. There are groups that "Re-Think Process," which is the way to explore change. The issue is how can a change in perspective be redesigned the way that the standard neoclassical economy paradigm works—designing products that can be 'made to be made' and the power of the system with renewable energy such as solar, wind, and hydroelectric. The process of rethinking starts with questions whether about basic ideas, creativity, and innovation, or some other areas so that people can build a strong reliable economy.

In contrast to the approach of minimization and dematerialization, the concept of ecoeffectiveness proposes the transformation of products and their associated material flows such that they form a supportive relationship with ecological systems and future economic growth. The goal is not to minimize the cradle-to-grave flow of materials, but to generate cyclical, cradle-to-cradle "metabolisms" that enable materials to maintain their status as resources and accumulate intelligence over time. This inherently generates a synergistic relationship between ecological and economic systems recoupling of relationship between economy and ecology.

Circular economy: from theory into practice

The Circular Economy is the answer to some of the main challenges of our time now, but especially for the near future due to escalation of climate changes. It helps today to preserve resources that are increasingly scarce and subject to greater than ever pressure on environmental areas ranging from natural resources to transportation, buildings, and wifi systems, plus more. Circular Economics has already boosted Europe's economy, competitiveness and environmentally "green" development along with smart green businesses.

Clark's book Climate Preservation (Elsevier Press, 2018) presents global cases (https://www.elsevier.com/books/climate-preservation-in-urban-communities-case-studies/clark/978-0-12-815920-0): that show how generating new business opportunities as well as innovative and more efficient ways of producing and consuming create businesses in communities that prevent and restore the climate.

As the world is round, so is Circular Economics, as it shows what happens in one part of the world will travel to other areas due to wind, rain, ocean water, and even earthquakes. The Circular Economy is global and therefore brings together local concerns to create opportunities for social integration and cohesion. Circular Economics even finds answers to the drastic fanatics to provide people with viable, safe and strong future for their families and children.

The paradigm transition toward a Circular Economy is the answer to some of the main challenges of our time. It can help to preserve resources that are increasingly scarce and subject to greater than ever environmental pressure. Initially, Circular Economics has already increased Europe's economy and competitiveness, by generating new business opportunities as well as innovative, more efficient, and new ways of producing, consuming, and creating new products. The chart later shows how the circular economy works. Moreover, it has been working in different countries and communities already.

Furthermore, it is recognized, at least in principle, that the only alternative to ensure a viable safe future for new generations of European citizens (and refugees) is that of a rapid transition to an industrial system that is based on the circular economy paradigm providing economic, family, and cultural support for everyone. Businesses and governments need to rethink production cycles so as to eliminate the concept of waste, through optimized models for the reuse of products, disassembly, and recycling of goods.

The circular patterns to which the industry should converge will have to specialize and focus on environmentally positive products, goods, and services to address the different issues related to fast moving consumer goods from durable goods as Global

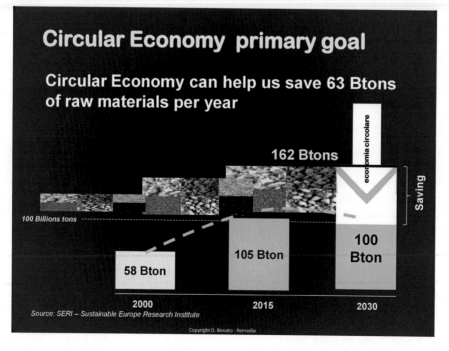

Sustainable Communities Handbook green technologies, public policies, and economics   (https://books.google.com/books?id=MNKjwmVYKN8C&source=gbs_similarbooks)

Circular Economics finds an answer to the terror of the fanatics: provide desperate people with viable, safe, and strong future for their families and children. Europe and the whole world are facing several pressing issues concerning climate change, the environment, society, and the economy, which are crucial to the quality of life of our children and grandchildren.

The European Union enacted circular economy

The European Union (EU) creation of a Circular Economy policy, program, and series of action plans was a long awaited package that plays a key role in supporting this transition to the future, today, by providing a clear message to the industry and society on the pathway forward. The package will have to drive investments and create a level playing field, removing obstacles stemming from European legislation, deepening the single market, and providing favorable conditions for innovation.

The "cold shower" on the expectations of those who had believed in this perspective came at the end of 2014, directly from the new European Commission established under the presidency of Jean-Claude Juncker, who decided to cut a long list of programs that he deemed ineffective or too expensive. Among those that he slashed is the program to establish, start, and institute the circular economy.

What exactly can the public and leadership expect? The experts who are working on the new package are well aware that, at present, about 90% of the European industry turnover (sadly one of the best performances in the world) is based upon linear economic models of procurement, production, consumption and disposal of matter. This neoclassical economic model is now defunct and proven to be only an ideological model that does not work in today's EU, let alone around the world for other nations. The chart later outlines how the Circular Economy could and is now working.

Meanwhile, the UN continues these meetings that will take place as it is critical to the future of all human beings and the earth. There are the reactions of the EU nations and the concerns of many Member States who have forced the EU Commission to do a review of its position.

Moreover, the UN will "restate" its earlier nonaction to implement a circular economy plan that makes solutions to climate change a reality and certainly provides the world with solutions that offset violence, wars, and terrorists attacks. The EU Commission had declared that the withdrawal of the first circular economy package was actually intended to define a "more insightful and strategic scope." Will this really take place? And especially now due to the need for world leaders to be secure and lead people to a new global reality to stop climate change with green smart communities that are no longer dependent upon fossils, their resources, and the economic forces behind them.

Certainly, the first proposal of the Commission was rather incomplete and focused primarily on waste management without addressing significant environment, climate, and economic areas.

(http://ec.europa.eu/environment/circular-economy/index_en.htm) For example, all other key aspects of this new economic paradigm, such as remanufacturing of things, sharing economy, and bioeconomy are significant to every nation and its citizens. Now, more than ever with millions of people from the middle-east all seeking survival and new lives, solutions are needed for climate change that provides support for them.

With this background in June 2015, a strategic EU conference was held, where European Union (EU) Commissioners Timmermans (First Vice President) and Katainen (Jobs, Growth, Investment and Competitiveness) solemnly promised to put determined efforts forward in support of the circular economy, calling it a "fundamental strategy" for Europe's development today. Therefore, in Brussels, the EU has started to work intensively to create a new and more comprehensive "Circular Economy Package," which should be ready by December 2, 2015.

Despite the November 2015 massacre in Paris, the clock is ticking and thus everyone will see this time what happens for the future. The expectations are high because investments in the circular economy will support the "green" (e.g., environmentally friendly) development of European industry and, at the same time, help achieve the goals that will emerge from the UN.

COP 21 on climate change in Paris will be held in early December 2015. As seen later, the circular economy differentiates from resource efficiency.

Circular economies in action

In the first case area on how Circular Economics works, the EU will push on the use of nontoxic natural ingredients, to be securely reintegrated into the biosphere. For durable goods, the goal is to ensure high rates of reuse and recycling, to avoid the hazardous substances that can harm the environment and to maximize resource efficiency. It is essential to achieve a strong decoupling of economic growth and raw materials usage from neoclassical economic model that never really existed.

To clarify the importance of decoupling, the public and governments all know now that the extraction of natural resources at European levels has increased from 12 billion tons in 1980 to the current 22 billion tons in 2015. Decoupling is therefore a strong requirement for the European economy, which would allow, among other things, to reduce the volatility in the prices of critical raw materials, giving stability to European industry and reducing supply risks.

In economic terms, the Commission estimates that, thanks to the Circular Economy, the savings in raw materials for industries by 2025 could be at least equal to 14% at the same level of output. Furthermore, this is a saving that could generate a benefit equal to

400 billion euros. In addition, the circular pattern would favor the creation of new business sectors, giving stimulus to consumption and promoting job growth.

Product and process design

One of the main tasks of the circular economy package is the development of innovative product requirements under the ecodesign directive, such as durability and recyclability. In this respect, it is very likely that the Commission will adopt a proposal to differentiate fees paid by producers in extended producer responsibility schemes according to the real end-of-life costs and recyclability of their products (http://ec.europa.eu/growth/industry/sustainability/ecodesign/index_en.htm).

As part of the regular reviews of best available techniques, the circular economy package should also include guidance on best waste management and resource efficiency practices for production processes in industrial sectors, improving the uptake of the European Eco-Management and Audit Scheme and the environmental technology verification system as well as methods to evaluate and decide on products (health, environment and nature ingredients, etc.) as "earth accounting" started to do within the circular economy paradigm.

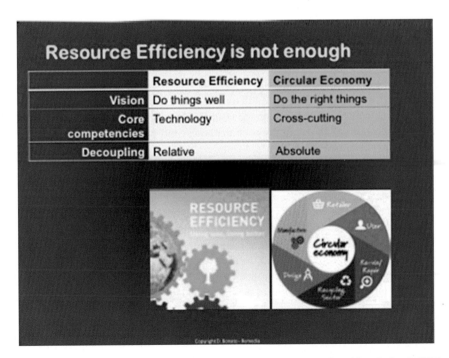

Copyright © D. Bonato 2015.

The Circular Economy package will also address in a systematic way the challenge related to the management of end-of-life products. Furthermore, the Circular Economy is one reason why Clark and Fast have been asked to update their book on Qualitative Economics (2008) (https://books.google.com/books/about/Qualitative_Economics.html?id=EPoSAQAAIAAJ) so that it is out from Springer Press in 2019, as the "invisible hand" never existed in the past and needs to be monitored for the future in a scientific manner. Governments have always been involved in nations and their development on all levels. "Civic or social capitalism" is needed now more today than ever before (http://www.amazon.com/The-Next-Economics-Environment-Climate/dp/1461449715)

Circular economy and waste management strategies

Each year in Europe, 2.7 billion tons of waste is generated, but only 40%—limited to a few streams—of this amount of waste is nonetheless collected and sent to reuse, recycling, energy recovery, or composting. Yet there is room for improvement, especially if we consider that in many European countries land filling is still the preferred option for waste management. Valuable but also hazardous waste streams are not properly tracked and managed along with illegally exported abroad. Even when recycled, the current processed are not designed to optimize the recovery of valuable raw materials. Case in point is the reuse of plastic materials:

Although the science of chemistry is important, the issue is to be able to recycle and reuse the plastic that is made and used for bottles, packaging, and more. Now, the plastic can be gathered from reusable products that provide a resource from which to make new products.

Moreover, collection systems are still too expensive and inefficient, which does not help industrial companies to abandon the traditional production systems based on the linear (flat economic) transformation of materials into products and their disposal once they are consumed. Therefore, the Commission is considering the possibility of introducing further simplifications to promote increased efficiency of collection systems through the circular economy paradigm. Hence, by integrating these systems with the upstream industries that make use of recycled components and raw materials from products entering, the end of their life stage can be profitably met as well as protecting the environment (http://ec.europa.eu/environment/waste/legislation/).

Another issue on which Brussels wants to put some focus is the development of professional networks specialized in the uses of equipment reconditioning and reusing products of all kinds. Thus, European nations can avoid the generation of waste and encouraging the development of new technical skills, and new jobs, especially for young people and their future tomorrow.

The new package will in some respect develop predemolition guidelines to ensure adequate recovery of valuable resources and proper management of hazardous waste, as well as voluntary industry-wide recycling protocols, based on the highest common standards in each waste stream. For example, one of many issues, waste management choices for nations in the EU. Variables and measurable policies are needed for waste management.

Standards for secondary raw materials

The Commission, through the circular economic package, will probably launch work (jobs, new businesses, and education) to develop quality standards for secondary raw materials where they are needed—in particular for plastics. The EU regulation on fertilizers will probably be revised, to facilitate recognition of organic and waste-based fertilizers, hence supporting the role of bionutrients in the circular economy. A key priority of the circular economic package will be the sharing of good practices between Member States and stakeholders on the cascading use of biomass and biobased products.

The Commission seems to understand that the demand generation from European industrial value chains connected to the waste management sector could be a clever, innovative, and environmentally friendly way that is best to feed a virtuous product production, distribution, and reuse circle based on the circular economy concept. It is no coincidence that in Europe, over the last 24 months, there are now several new companies created and operating in the field of secondary raw materials brokerage services to efficiently connect supply and demand generating benefits for the market and significant profits for investors.

Among the other key areas, the Circular economy package will also develop analysis and policy options to facilitate shipment of secondary raw material across the EU (electronic data exchange and possible other measures), and improve data availability on raw materials. Such standards are measurable and thus able to be evaluated, if needed for changes, revisions, and improvements. The results are then in other areas where the circular economy can be developed, implemented and measured based on these results. Thanks to these initiatives, experts expect that the circular economy package will enable more effective strategies to increase the demand for secondary raw material coming from reuse and recycling operations (http://www.rreuse.org).

In this perspective, the new circular economic package aims to encourage the creation of new industrial initiatives, based upon a greater integration between different companies and businesses along with governmental institutions. By deploying in a more effective environmental way, the most innovative information and communications technologies reduce information asymmetries that hinder trade, but develop ground-breaking cooperative processes.

The Circular Economy package also addresses in a systematic way the challenge related to the management of end-of-life products. Each year in Europe, 2.7 billion tons of waste is generated, but only 40%—limited to a few streams—of this amount of waste is nonetheless collected and sent to reuse, recycling, energy recovery, or composting. Yet, there is room for improvement, especially if we consider that in many European countries land filling is still the preferred option for waste management. Valuable but also hazardous waste streams are not properly tracked and managed along with illegally exported abroad. Even when recycled, the current processes are not designed to optimize the recovery of valuable raw materials.

Nations need a solid systemic ecoinnovation, inserted in a broader perspective, that considers all the innovative solutions able to contribute to the development of the three fundamental "axes" of sustainability: economic, social, and environmental. From this point of view, the circular economy is the most effective approach in the model of development and has recently become the strategic reference for all the countries of the European Union and the commitment of the Belgian Presidency and Council of Union European Union (Flemish Environment Minister Joke Schauvliege). Below shows circular economy impact land, water, and air: https://www.ellenmacarthurfoundation.org/

The reference models underlying the circular economy stem from the collaboration between the industrial design strategist, William McDonough (www.cradletocradle.com) and the German chemist Michael Braungart, both collaborators of the Ellen MacArthur Foundation.

One of the key priorities for Europe in the Circular Economy is that of resource efficiency and waste reduction (Zero Waste Strategy). In this area several *stakeholders* are active such as (http://www.consorzioremedia.it/), the collective system of manufacturers of electrical and electronic equipment that is developing a long-term vision to fully exploit the potential of secondary raw materials and to increase innovation capacity of the recycling sector in the EU.

The dynamics linked to the Circular Economy can thus be transformed into a solid pillar of the growth strategy of the European Union. The world of industry, in collaboration with governments, must work to address social environmental challenges and generate concrete benefits for individuals and their communities. What is of particular importance and reflected in the book is the need for qualitative economics to be an active part of the solutions to climate change as well as in creating new businesses, industries, and areas where products can be recycled.

So far in the book, there are some examples of companies such as thin film solar from Hanergy

(http://www.hanergythinfilmpower.com); made into a new product such as clothing; provided reuse of components in media, computer, and other tech systems (e.g., as Remedia Company does (www.consorzioremedia.it); or even the reuse of waste paper into business cars and reusable paper that can be planted to grow flowers (www.4imprint.com/nelson/art.htm) and there are more and more coming around the world and doing business globally. For example, Hanergy has opened in early 2018 (http://www.hanergyamerica.com) an office in Silicon Valley that ONLY sells thin film solar. This office covers all of the Western Hemisphere so the business will expand in the United States and other nations.

Smart Green small and central Power Grid

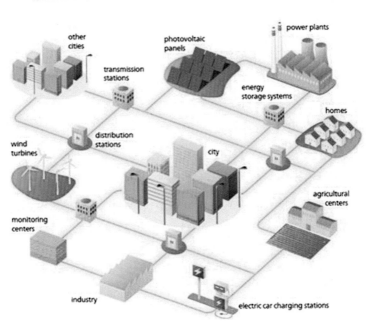

From Lixuan Hong, PhD. Thesis at University of California, Berkeley

The qualitative economic strategy can only be made if we know how to acquire in-depth knowledge about the availability of secondary raw materials. It is also necessary that the actors of the system (raw materials industry, end users, institutions, companies, and consumers) collaborate closely to achieve common reuse and recycling targets along the entire production and distribution chain. In this way, consumers will know exactly what they are buying and the impacts and potential for reuse and recycling of the products they have chosen.

There is also an absolute need to improve the awareness of European industry in access to critical raw materials, thanks to a closer monitoring of the global markets of nonenergy resources. Too many companies ignore the fact that their products are threatened by the risk of future, reliable, and fair supply of strategic resources. Below are the economic importance of numbers and statistics that also need to be defined, verified, and repeated.

Critical raw materials matrix of the European Commission. This is a quantitative methodology that applies two criteria—the economic importance and the risk of procurement of selected raw materials. Economic importance: this analysis is obtained by evaluating the share of each material associated with macro industrial sectors at European level. These quotas are therefore related to the gross value added of the macro sectors compared to the GDP of the EU. The value obtained is then evaluated on the basis of the total EU GDP to define an index of global economic importance for a given material.

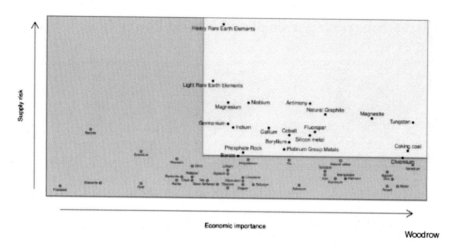

Economic importance

Woodrow

Supply risk: the World governance indicator is used to measure this risk. This indicator takes into account a very wide variety of criteria, such as levels of responsibility, political stability, absence of violence, government effectiveness, quality of legislation, rule of law, and control of corruption.

Minerals and critical metals are essential for environmental technologies such as solar photovoltaics, wind power, and lighting, and for low-carbon electric vehicle industries. Interruptions in the supply chain in times of crisis can be dramatic from the environmental, social, and economic point of view. This is the main reason why there is a need for a global framework in access to raw materials, capable of overcoming the traditional neoclassical model of supply and demand.

Many modern products, especially those of high-tech media and those that transit the web and social networks are truly global, from their creation to distribution and sales. Tools like Netflix, YouTube, Facebook, and other forms of communication must be part of a circular economy model that directs toward the reuse and recycling of products to create new ones. The whole process can and must be improved.

Success will only be seen when we are able to achieve a series of improvements over access to and availability of raw materials, with sustainable resource management within a greener, more circular economy and global ecoinnovations. In technology both in commercial practices to meet the growing demand and ensure the needs in raw material supply. This effort should also lead to changes in behavior that are more consistent with consumers' sustainable use of raw materials.

Moreover, collection systems are still too expensive and inefficient that does not help industrial companies to abandon the traditional production systems based on the linear (flat economic) transformation of materials into products and their disposal once they are consumed. Therefore, the Commission is considering the possibility of introducing further simplifications to promote increased efficiency of collection systems through the Circular Economy paradigm. Hence, by integrating these systems with the upstream industries that make use of recycled components and raw materials from products entering, the end of their life stage can be profitably met as well as protecting the environment (http://ec.europa.eu/environment/waste/legislation/).

Another issue on which Brussels wants to put some focus is the development of professional networks specialized in the uses of equipment reconditioning and reusing products of all kinds. Thus, European nations can avoid the generation of waste and encouraging the development of new technical skills, and new jobs, especially for young people and their future tomorrow.

The new package will in some respect develop predemolition guidelines to ensure adequate recovery of valuable resources and proper management of hazardous waste, as well as voluntary industry-wide recycling protocols, based on the highest common standards in each waste stream. For example, one of many issues, waste management choices for nations in the EU.

European funds for innovation and skills development

Europe and the whole world are facing several pressing issues concerning climate change, the environment, society, and the economy, which are crucial to the quality of life of children and future generations.

The European strategy for the circular economy will rely on Horizon 2020, the mainstream—80 billion euros innovation program activated by the EU Commission. Through Horizon 2020, several high impact European business value chains will speed up their transition to the circular economy (http://ec.europa.eu/programmes/horizon2020/). Hence, it is worth emphasizing that Europe present several cultural and infrastructural strengths to be exploited, which could place the entire EU into the global leading position toward the circular economy.

Europe can lead and is doing so now (2019) in sustainable mobility, remanufacturing, sustainable development of nature-based solutions, and implementation of new hydrometallurgical processes for the recovery of rare earths and precious metals.

Furthermore, it would be important to support industrial companies in developing a clear vision of their priorities, by choosing few selected high impact projects to be targeted and thus funded. Value chain leaders are required, with the capability to aggregate different partners (including industry, research centers, academia, institutions, governments, etc) and lead them toward the realization of a new industrial model based on the circular economy paradigm.

The circular economy paradigm certainly addresses the goal of developing a stronger innovation culture, by supporting greater action in education, both at academic and at industrial levels. As already mentioned, the EU Commission is aware that the lack of skills and professionalism in institutions and companies is one of the biggest obstacles in the adoption of a new industrial approach.

The priority for education and jobs must begin immediately and build upon programs, publications, and people who are already involved in the Circular Economy.

Simplify to innovate

Needless to say, several European countries, such as Sweden, UK, and France, are already strongly involved with the circular economy to modernize and simplify the current legislative framework needed from the European Commission. In modern economies, the environment is an essential resource to be protected, but today not only and so much through a formal, prescriptive approach.

Rather, it should be possible to precisely measure the impact of environmental externalities ranging from environmental costs associated with the use of ecosystems by individuals and businesses (http://www.amazon.com/The-Next-Economics-Environment-Climate/dp/1461449715) that are penalizing those who do not change their bad habits and rewarding those who design the company business with the goal of mitigating their impact on the ecosystems ( ).

Today, very often, these attempts with public policies and economic programs are thwarted by rigid and outdated legislation, which sees waste solely as an environmental issue and not as an opportunity to create any economic and societal values. Yet, as the graphic later illustrates, all of these issues and their solutions are integrated and need to done.

The European Union has already changed

Some of the more advanced European industrial companies have anticipated the Commission moves in Brussels. Many nations have begun a transition of their production systems and service to the new Circular Economy structure. This is happening because the most innovative (http://www.ellenmacarthurfoundation.org/news/towards-the-circular-economy) enterprises have realized that the adoption of production models based on circular economy favors the development of their business and create shareholders' value.

The most useful strategies range from remanufacturing of exhausted products to the transformation of products into services, according to the paradigm of the circular economy for sharing, from zero-waste policies to supply chain redesign aimed at implementing new symbiotic partnership. This is already a reality in the European electronics industry, where manufacturers have created dedicated organizations focused on resource efficiency, recycling, and ecodesign such as Remedia in Italy (http://www.consorzioremedia.it/), Ecosystemes in France (http://www.eco-systemes.fr/), and Repic in UK (http://www.repic.co.uk/).

The European Commission has expressed its intention to encourage and accelerate industrial investment for the adoption of Circular Economy paradigm models, rewarding the most worthy programs and ideas. Tracking and watching these intentions will translate into concrete measures, capable of transforming the current experimental stage in a so-called mainstream scenario. Hence, the industry reference models will actually be based on the Circular Economy paradigm as will be noted later.

In the fall of 2018, Joss Blériot, the Head of Public Affairs at the Ellen MacArthur Foundation since 2012 noted what Circular Economics is in the EU and but soon other nations, like China. From a European perspective, it is clear that the EU Commission is leading the world in Circular Economics as it is seeing how the principle of the circular economy package has advanced since it was adopted in 2014. At the time, not one member state had come up with a national roadmap, and it was very much the hope of the Commission that the nations would in any case start Circular Economics to implement it sooner than later.

What is seen as a result now is Circular Economics is being used and spread across the EU and into other Continents. For example, Finland presented its roadmap in 2017. France has just published one also while Slovenia published its own strategy in May 2018. Italy has revealed the building blocks of its roadmap as will be outlined later. The Netherlands were advanced before the Circular Economy package was announced, yet it was not a fully-fledged national roadmap. For them, the idea (now with a name, Circular Economics) became a reality as the nation combined and strategized for decades on their wind power from the ocean as well as energy from run-of-the-river power from rivers.

Moreover, Germany has been working on bits of a circular economy, although their definition is closer to resource efficiency, which is an area that they have been involved in for a long time due to their dependency on coal that has stopped. However, Germany is now dependent upon natural gas from Russia as many other EU nations have done also. So there is the convergence of signals from a European perspective that the impetus for Circular Economics was given by the Commission. Then, it spread to EU member states.

The business, industrial, and corporate strategies saw the value of Circular Economics, so they were very active and focused on promoting and enacting the decisions from the Commission in Brussels. The advantages grew so now the EU has created and implemented and economic set of areas that have not only been inexpensive, but also protected the environment, reduced climate change, and created new jobs, work, and even education focused on implementing Circular Economics.

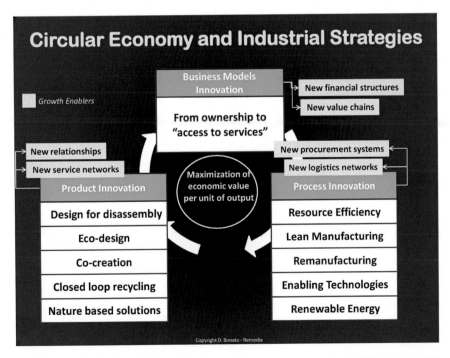

To be clear, few people (in government, academics, and even business) know what Circular Economics is all about. In European, nations and communities see the need for an effective set of measures to support the adoption of their circular economy as an essential component for the EU growth strategy. Most of the required ingredients are already in place and they are ready to go. However, first of all, the European Commission needs to deal with the cultural and innovation gap at all levels ranging from consumers and their communities to industry and business to public institutions, government, schools, and society itself.

The "Circular Economy" in Europe

The first priority for the new Circular Economics package was to build and leverage a favorable government policy and plan for the development of new ecoinnovative and systemic industrial models inspired by the circular economy paradigm.

The long awaited circular economy package from the European Commission would therefore provide the necessary tools to address and overcome the current and future limits. Stakeholders expect effective measures to promote and reward companies that invest in the design of products, services, and industrial processes based on a systemic, ecoinnovative vision, with a strong focus on Small Manufacturing Enterprises (SMEs).

To accomplish this critical end result, it is important to identify the leaders, study their strategies, and provide effective and balanced rewarding mechanisms (ranging from economic to environmental and modernization). Such incentives should be aimed at companies that are committed to transform products into services, while maintaining ownership of the goods sold to more effectively manage their end use.

Financial service is an important civic capitalism strategy to provide better and tailored answers to consumers' needs. In addition, new measures to stimulate the circular economy should fit into a broader framework related to the rethinking of the current fiscal policies, in support of green economy and ecoinnovation that protects our environment and air for future generations.

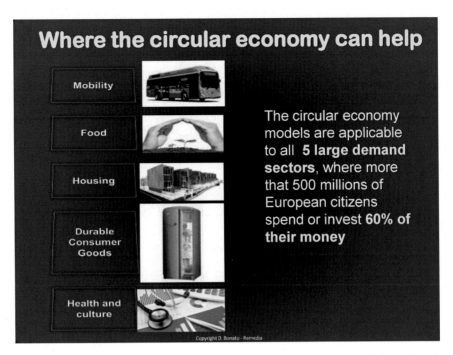

Several European countries, such as Sweden, UK, and France, are already strongly involved with the Circular Economy to modernize and simplify the current legislative framework needed from the European Commission. In modern economies, the environment is an essential resource to be protected, but today not only and so much through a formal, prescriptive approach.

Rather, it should be possible to precisely measure the impact of environmental externalities ranging from environmental costs associated with the use of ecosystems by individuals and businesses that are penalizing those who do not change their bad habits and rewarding those who design the company business with the goal of mitigating their impact on the ecosystems.

Today very often these attempts are thwarted by rigid and outdated legislation, which sees waste solely as an environmental issue and not as an opportunity to create any economic and societal values. Yet as the graphic later illustrates, all of these issues and their solutions are integrated.

Case from EU: Italy

The two organizations have also signed a Manifesto for the Circular Economy, along with a group of businesses from different industries including materials company Novamont, consumer goods group Fater, and the food marketplace Eataly. The aim of the manifesto is to promote collaboration and knowledge sharing, to support the development of a circular economy outlined in the paper from the government.

The paper had been published by the Italian Ministry of the Environment and Ministry of Economic Development, and seeks to provide an overview of the circular economy and clarify the strategic positioning of Italy on the topic. It also links the circular economy with other commitments that are on the agenda of not just the Italian government, but policymakers and industries around the world, such as the Paris Agreement, UN Sustainable Development agenda, and the European Union.

It builds on existing momentum toward the circular economy in the country. Italy held the 2017 presidency of the G7, and has used this role to push forward activities that encourage better use of resources and energy, and the elimination of waste and pollution. In June of this year, the G7 launched the Bologna roadmap that identified common priority areas for group members, including criteria for extending the life of products, tackling food waste, improving plastic use, and the digitization of production.

In a report "Towards a circular economy model for Italy," the authors claim that the Italian economy is indeed becoming more "circular," due to increasing waste separation and a reduction in overall use of resources.

The time to shift from the theory of "need" to an actual "opportunity" is in the Circular Economy by leveraging Italian technological capabilities and its momentum in Europe. This means redesigning products and business models, looking for new connections through industrial symbiosis, creating a regenerative bio-economy, and moving toward a model of product responsibilities. Moreover, above all, the need to offer new financing models to support the "circular innovation" has become a top national and local priority.

Conclusion

In summary, Circular Economics uses both quantitative and qualitative data for everything from newCos to old companies to wifi, nonprofits, and governments around the world the point is that while Circular Economics may sound vague and even difficult to define, it is not. As noted earlier and through the book, the world (and economics) is round; not flat.

The Ellen MacArthur Foundation started the research and work in Circular Economics that has now turned into the Economic Policy and Plans for the European Union, and now going to be started in China too. There are other nations doing the same. The United States has NOT gotten there yet.

However, the USA will soon be enacting Circular Economics due to the CE staff and collaborators who are making plans for CE for conferences, key players and businesses. Qualitative Economics plays a key role in Circular Economics, as there is a need to be scientific in the definitions of terms, numbers, data, and everything. People need to define what they are doing and want to do so that others can join or even reject them. Hence, the science of linguistics is noted with a graphic from Noam Chomsky:

What qualitative economics wants to do is define numbers but also works, sentences, and even pages. The key is to explain things so that people and organizations

Linguistic Transformation Theory
(N. Chomsky 1975)

```
T  Surface Structures (Phonetic- Everyday Language)
R<------------Language Discourse ----------------->
   A        Universal Grammar              ^
   N        Syntax                         |
   S          Data (methodology: interactive/qualitative)
   F        Empirical (actual use of language) |
   O                                        |
   R        Deep Structures (Semantics –    |
   M        meaning to words/sentences)     |
   A                                        |
   T        Generative                      |
   I        Phrase Markers                  |
   O   Rules (principles that form language): |
   N        Appropriateness etc.            |
   S        Lexicon                         v
   <------------Definitions (understanding) ------------->
```

understand each other. Above all, Qualitative Economics is the key component to making economics into a science. Quantitative data is not enough information, data, and understanding for anything. Moreover, as noted in the earlier chapter, it can be very deceitful on purpose to enhance a project; make money; or even rob people, companies, and governments.

Some of the work in qualitative economics is now from a legal perspective, not just one that is either in economics or another social "science" such as politics, sociology, and anthropology. Above all, the science of economics needs to be interdisciplinary, as very few anthropologists know what politicians are talking about. That is why, there are deep structures in language that explain the surface structure about what is being said, why, and above all the next steps. Linguistics has gotten in this area due to Chomsky and now more and more legal educators are using this kind of explanation about words, data, and even formulas.

The Next Economic Paradigm is both Qualitative and Quantitative that has stimulated a lot research and more publications. The need to be Cross or Interdisciplinary needs to be applied to everyone's life be it in family, business, or government. As economics is not linear, but circular all sectors, economic growth, and careers/jobs are impacted. Moreover, there are very positive ways to use qualitative economics in Circular Economies, as everything in our Lifeworld is or gets connected. That is why, recycle and reuse of products is so important. The use even of raw materials is significant as it needs to be reduced to pressure our local and global ecosystems.

The point of using the case of electric cars is important globally today. Yet tomorrow (actually now) there are all solar-powered cars. Energy systems are critical in humanity from living to working with technologies, policy, and economic solutions that are good for the environment.

Case in point noted is Hanergy, a company that produces and manufactures thin film solar. Here is their product that is created and reused in many ways:

Here is their use of thin film solar for the buildings at Hanergy Group Headquarters in Beijing where they are located on the Park where the Olympic games were played in August 2008. Hanergy has eight (8) buildings that are all using thin film solar: 43% of their energy for all the buildings comes from thin film solar. In the context of Circular Economics, Hanergy has since 2016 created all solar-powered cars and a bus.

Clark has been in one solar car when they were released in July 2015 and two more after that in 2016 and 2017. The change in economics into becoming a science means that almost everything needs to be reconsidered into a thinking process that gets people, groups, companies, and government to explore change. Not change just in statistics and charts but in actual definitions and explanations about the changes in a science manner.

Acknowledgment

Special thanks to Ellen MacArthur Foundation. Circular Economics. Programs. The Circular Economy Opportunity for Urban and Industrial Innovation in China. ARUP Report, December 2018.

Planning for more sustainable development

9

Akima Cornell, PhD

Principal, Akima Consulting, LLC, Los Angeles, CA, United States

Overview

International and local climate-change policies and regulations are compelling cities to change the way they plan and operate. Cities are now developing high-level planning to develop a strategic approach to become more sustainable and reduce greenhouse gas emissions from multiple sectors including energy, water, transportation, and waste. Developing a strategic plan is a key step in developing sustainability programs to reduce GHG emissions. Sustainability plans appeal to cities as they present a balanced approach to the environmental, social, and economic needs of their areas. These items include environmental, social, and economic impacts and are not limited to issues that have a significant financial impact on a city. Cities need to determine what aspects are the most material for their development, management, and growth for the present day and the near future. This chapter discusses how cities are developing sustainability plans and what their process is for developing goals and targets.

Why are mega cities developing sustainable plans?

Cities view sustainability as creating "development that meets the needs of the present without compromising the ability of future generations to meet their own needs."[1] As a result sustainability planning is seen as critical to the future development and operation of any major city. As sustainability planning is incorporated into the fabric of a city the definition and application of sustainability has evolved to become a balance between the environmental, economic, and social needs of a city. The balance between these three pillars is necessary for true sustainable development, that the short-term gains in one pillar may have drastic impacts on the others.

Author John Elkington coined the term triple bottom line; wherein organizations needed look beyond building profits, a single bottom line, to create long-term

---

[1] Development: Our Common Future (World Commission on Environment and Development (WCED) (1987) Report of the World Commission on Environment and Brundtland Report). Chapter 2, pp. 4

Sustainable Mega City Communities. https://doi.org/10.1016/B978-0-12-818793-7.00009-3

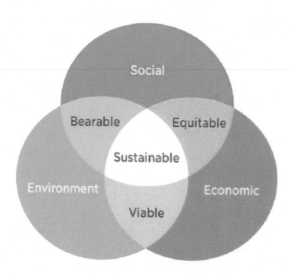

Triple Bottom-Line, Thomopolos & Embry (2013).

benefits for the environments and communities in which the organizations operate.[2] Nikolas Thomopoulos and John Embry expanded upon this construct noting the overlap between the three pillars can provide a more beneficial outcome, however to be truly sustainable there must be an equitable balance between all three pillars.[3] Similarly, cities are looking to develop and grow, but in a way that maintains resources and generates opportunities for generations to come. As a result, cities are developing sustainability plans that are designed to guide long-term development, programs, and resource allocation.

The process for developing a sustainability plan is as varied as the cities that create and implement them; one size does not fit all. However, there are a number of key steps in the process that most planning committees adhere to develop their plans. The process involves ascertaining the existing resources, projects, services, and gaps—this is establishing a baseline. The baseline gives a city a starting point, what are areas in which the city excels, and what are areas in need of improvement or are vulnerable to external impacts such as natural disasters or economic downturns? Some cities take things a bit further and evaluate the success and drawbacks of other

[2] Elkington, John, 1997. Cannibals with Forks: Triple Bottom Line of the 21st Century Business, New Society Publishers (reprint).

[3] Thomopoulos, Nikolas & Embery, John. (2013) *Two birds with one stone: enhancing education for sustainable development and employability* © 2013 Advance HE.

cities. This is benchmarking—the "keeping up with the Joneses" of city planning.[4] Additionally, cities have begun to engage stakeholders, key partners, and to a certain extent the public at large to provide input and feedback on the development of the sustainability plan, an aspect not commonly found in other areas of city planning. Once the plan is drafted and evaluated by stakeholders and city officials the plan is rolled out to the public and the city at large for implementation.

Baselining and benchmarking

Baselining

The sustainability planning process begins with establishing a baseline understanding of the existing resources, projects, and policies as well as identifying gaps and major sustainability challenges within the city to identify the key issues that will be addressed in the plan. Creating a baseline is also critical for establishing a starting point to measure performance and progress toward meeting goals and targets. A baseline assessment provides an inventory and review of current sustainability performance as determined by its related activities, policies, and procedures.

Indeed, Johnson et al. describe sustainability planning a 5-step cyclical process, in which step 1 is developing a baseline assessment for which "… there are two purposes for conducting this assessment: (1) to determine which sustainability factors, if any, need attention in the planning and implementation steps and (2) to provide baseline data for evaluating the impact of the sustainability actions."[5] Creating a baseline is an important first step in the development of a city's long-term sustainability strategy that will support the city's economic vitality, ensure the efficient use of resources, reduce environmental impacts, and enhance the social benefits to the community.

Baseline information and data are gathered to capture current levels of performance, showing city achievements and gaps in service. Cities engage their department to develop strategies to meet the sustainability goals that correspond to their areas. The process of establishing a baseline requires a combination of conducting data collection through inventories, resources assessments, surveys, and interviews. The baseline creates actual and current data on the existing projects and resources, creating a status report before developing and implementing a sustainability plan.

Cities vary in the type and depth of information gathered to create a baseline. The majority of cities gather information on their energy, water, waste, renewable energy, buildings, and transit services to ascertain their baseline carbon footprint.

[4] Momand, Arthur R., 1913. "Keeping Up with the Joneses" Comic the New York World.
[5] Johnson, Knowlton, Carol Hays, Hayden Center, Charlotte Daley, 2004. Building capacity and sustainable prevention innovations: a sustainability planning model. Evaluation and Program Planning, vol. 27. Elsevier, pp. 145−6.

Each of the areas becomes a key aspect of the sustainability plan. For example, a city could ascertain part of its energy usage through building utility bills. This would provide the city with an initial amount of energy consumed by its buildings. Once the city began implementing energy reduction measures, such as switching to energy efficient lighting fixtures and appliances, the city would then be able to ascertain energy reductions and the effectiveness of the project. As a result, baselining not only creates the starting point to aid cities in developing their sustainability plans, but also provides the initial metrics to measure the progress toward goals.

Benchmark

Benchmarking enables a city to measure, through existing and new metrics, its overall sustainability performance in comparison to other cities. Researching and collecting data on the cities of comparable size, area, and economy provide an important point of comparison. Reviewing other cities' sustainability plans provides a roadmap to what sustainability goals, targets, and strategies are feasible to undertake. A city can also take advantage of best practices and lessons learned from developing and implementing a sustainability plan.

Cities can also adhere to government standards, international guidelines, industry certifications, or best practices. As Boyko et al. explain, "Benchmarks could be current guidelines (i.e., what is required), current best practice (i.e., what is achievable), current typical practice or behaviour (i.e., what is currently done), or the current situation (i.e., what is happening now as a result of what has happened in the past; an average)."[6] Examples of industry certifications or best practices include US Green Building Council LEED certification, International Standards Organization, Green Building Council of Australia, and Comprehensive Assessment System for Built Environment Efficiency in Japan, which provides guidelines on sustainability building codes for cities to model their own plans and codes.[7] Additionally, cities can model their sustainability plans based on the United Nations (UN) Sustainable Development Goals.[8] Established in 2015, the UN Sustainability Development Goals were developed in coordination with governments and businesses, setting 17 overall goals and 169 targets and a roadmap for creating a sustainability plan.

---

[6] Boyko, C.T., et al., 2011. Benchmarking sustainability in cities: The role of indicators and future scenarios. Global Environemtal Change, https://doi.org/10.1016/j.gloenvcha.2011.10.004.

[7] US Green Building Council LEED (https://new.usgbc.org/), International Standards Organization (https://www.iso.org/standards.html), Green Building Council of Australia (https://new.gbca.org.au/green-star/), Comprehensive Assessment System for Built Environment Efficiency in Japan (http://www.ibec.or.jp/CASBEE/english/).

[8] United Nations Sustainability Goals, website: https://www.un.org/sustainabledevelopment/sustainable-development-goals/.

United Nations Sustainable Development Goals.

Source: https://sustainabledevelopment.un.org/?menu=1300.

Cities create their benchmarking reports by aligning the key indicators or areas of interest (e.g., energy, water, waste, renewables, transportation, etc.) noted from the baseline assessment. Cities can compare sustainability goals, targets, strategies, standards, and practices across multiple sources to establish their own plan.

Outreach and engagement

Developing an outreach plan to engage stakeholder can provide important legitimization and public buy-in for cities developing a sustainability plan. Stakeholder outreach during the city planning process is important to maintain broad community understanding and support. Actively engaging the communities and stakeholders in discussing aspects of the sustainability plan helps to build a collaborative open dialogue among the city, the public, and key stakeholders. This process could involve meeting with local civic and business organizations, holding public open houses and focus groups, or developing online surveys and meetings. Working with a broad range of stakeholders is important for identifying and engaging community leaders who often become spokespeople, letter writers, and campaign advocates as the process of developing a sustainability plan moves forward. Stakeholders groups and organizational leaders typically also find key project concerns early on that can benefit from discussions and mitigation.

Cities often hold outreach meetings and develop education materials in multiple languages to engage the communities. Utilizing personnel who are fluent in multiple languages are better equipped to provide outreach and education to a diverse population. Another major component in which cities develop and implement public outreach is to establish and maintain a website that provides up-to-date project information, a venue for submitting and tracking comments, a calendar of outreach activities, and opportunities to join an outreach e-mail list. Websites can also support interactive functions, including e-mail blasts, and online surveys. Cities are also including techniques for engaging the community through social media, including Facebook, LinkedIn, Instagram, and Twitter.

To ensure community and stakeholder feedback is incorporated into the plan, city and supporting staff must devise a means of documenting and cataloging the questions and comments. Online platforms, such as the city websites and social media, will provide a time-stamped record of public comments. To document in-person meetings and interviews with stakeholder groups, staff members will need to develop comprehensive notes to incorporate comments into the plan.

In addition to overarching public outreach, cities may elect to perform more targeted outreach, particularly to contact low income, minority, and disenfranchised populations. Cities will also engage local businesses and community organizations during plan development to ensure that the goals and strategies that emerge from the discussion support existing projects and services.

Drafting and implementation

Once a city has completed the community outreach and stakeholder engagement phase, the city and supporting staff can begin developing the draft sustainability plan. The baselining and benchmarking will have defined the sustainability areas of focus that become the chapters of the plan (e.g., water conservation, green jobs, land use, homelessness, etc.). Information from the engagement efforts should be used to guide the drafting of the plan to address public questions and concerns. The sustainability plan should reflect the city's sustainability priorities and include an overarching vision for the city. Chapters outline city goals and targets for a specific sustainability topic, as well as action items or strategies for implementation.

Developing the sustainability plan should include an evaluation component. Measuring the effectiveness of the sustainability plan is borne out when measurement and evaluation are considered part of a larger iterative process with monitoring and reporting being embedded in the implementation of the plan.

Evaluation includes clearly identifying the targets, implementation strategies, project objectives in measurable terms, identifying key performance indicators, outlining data collection and analysis methods, and developing a timeline to monitor the success of the program overtime. Setting up these evaluation procedures during the plan's inception is key to implementation and documentation moving forward. The city can improve implementation and monitoring by creating an evaluation team or unit to collect data and monitor progress toward goals. The team should have access to all departments and data streams as well as have the authority to present accurate if occasionally controversial findings and be protected from unforeseen problems. The evaluation team should have the ability to provide changes in project implementation and reallocation of resources and staff.

Case studies
City of Milwaukee

ReFresh Milwaukee.

Source: https://itmdapps.milwaukee.gov/citygov/refreshmke/index.html.

In July 2013, the City of Milwaukee published its first sustainability plan "ReFresh Milwaukee" with guidance from the mayor "to build a smarter city through sustainability."[9] The plan outlined how to foster Milwaukee's economic development by building stronger, more resilient neighborhoods, and focus on building environmental benefits. The plan recognizes the overlapping benefits to neighborhoods and residents benefit from building a plan that provides clean air, water, and green spaces; adequate transit; healthy, affordable, and culturally appropriate food; reliable and renewable energy; and safe, efficient infrastructure. In addition

---

[9] ReFresh Milwaukee, 2013. City of Milwaukee Sustainability Plan: https://city.milwaukee.gov/ReFreshMKE_PlanFinal_Web.pdf.

to environmental benefits, Milwaukee's sustainability plan also addressed the social and economic needs of the city including human capital, health, safety, education, and equity of all its citizens.

In 2012, the Mayor of Milwaukee commissioned a Green Team, comprising city staff from the Environmental Collaboration Office (ECO), representatives from neighborhood organizations, nonprofits, business and civic organizations, and expert sustainability consultants to develop the sustainability plan and commit to making social equity and environmental justice a comprehensive lens in the envisioning all goals, targets, and strategies of this plan. The plan was intended not only to address the long-term sustainable growth, but also to be a catalyst for change and revitalization for Milwaukee's neighborhoods.

The Green Team was tasked with developing a sustainability plan that "creates an alignment of economic and environmental interests that improve Milwaukee's quality of life both for current residents and businesses and for future generations through embracing smart, achievable sustainability principles."[10] The plan was designed to guide Milwaukee through the next 10 years, a timeframe that reflected the long-term approach and complexity of implementing the plan. An important aspect of the plan was that Milwaukeeans created it as a bottom-up approach to planning and development.

To ensure that the efforts Milwaukee's sustainability plan reflected the priorities of residents, Mayor Barrett and the Green Team reached out directly to residents through town hall meetings, smaller group discussions, and a bilingual public survey to help identify the issues in which to focus. In 2012, the Green Team surveyed thousands residents in an initial round of public outreach that gathered input and direction on the issues and priorities to address. In addition, the Green Team hosted 5 formal town halls and more than 30 smaller focus groups that reached an estimated additional 400 residents, bolstering the survey input with in-person contributions. The Green Team also worked with area businesses and several trade groups to gather direct feedback from Milwaukee's business community, with approximately 85 businesses participating.

The online survey had eight questions that outlined the guidance and priority setting of residents and were backed up by in-person feedback. Although residents agreed that sustainability is an important concept to guide the city now and into the future, everyday issues must be addressed first to create a foundation for sustainability. These issues include access to jobs, quality educational opportunities, and the need for safer neighborhoods.

Surveys and interviews also confirmed that residents view diversity as beneficial and emphasized the importance of creating connections between neighborhoods. Residents also identified other important challenges and concerns, including empty lots and abandoned buildings, public transportation, neighborhood crime and graffiti, and access to healthy food.

---

[10] ReFresh Milwaukee, 2013. City of Milwaukee Sustainability Plan: https://city.milwaukee.gov/ReFreshMKE_PlanFinal_Web.pdf.

The Green Team used this information to guide development of the sustainability plan and craft goals, targets, and strategies to help residents, businesses, and other Milwaukee organizations improve the neighborhoods. The goals were organized into eight main chapters that were derived directly from residents' input. These themes were buildings, energy, food systems, human capital, land and urban ecosystems, mobility, resource recovery, and water. The goals and targets within each of the chapters provides a strategic framework, or roadmap, for encouraging individual responsibility, city leadership, and purposeful action that will ultimately lead to collective benefits for Milwaukee.

Implementation

City of Milwaukee ECO logo.

Source: https://city.milwaukee.gov/eco#.XPCt59PYqi4.

Implementing the sustainability plan as a whole entails a separate set of roles, responsibilities, and resources. First and foremost, the city needs to track Milwaukee's progress toward the sustainability goals and targets outlined in the sections of the sustainability plan, report progress to the community and concerned stakeholders, and identify ways to continue improving sustainability performance. Ideally, the various individuals, city departments, agencies, and organizations would work together to implement the plan strategies will coordinate their efforts.

The city's ECO Department was responsible for overall sustainability plan implementation, including annual evaluation, reporting, and public engagement. ECO would maintain a voluntary Green Team, which included current members and add new members as appropriate, to serve as a coordinating body. City of Milwaukee Green Team members and ECO staff served as the official leads for the eight base conditions (buildings, energy, food systems, human capital, land and urban ecosystems, mobility, resource recovery, and water). These individuals coordinated directly with the relevant city departments and community stakeholders on strategy development and implementation and data tracking.

City of Los Angeles

The City of Los Angeles released its "Sustainability City pLAn" in April 2015, to act as a roadmap for near-term and long-term sustainable progress over the next 20 years. The plan was built on the three pillars of sustainability—environment, economy, and equity. These pillars require a balanced approach to the needs of people as well as the environment. Elkington referred to it as the triple bottom line, denoting that organizations needed to not only care about their financial bottom line (economy), but also their impacts on the surrounding stakeholders (equity) and area in which they operate (environment).[11] To operate in a manner that supports all three aspects would benefit the company in the long-term. As a result, Mayor Garcetti focused on building a plan that was "comprehensive and actionable directive that will produce meaningful results for today's Angelenos while setting us on the path to strengthen and transform our city in the decades to come .... by addressing the environment, economy, and equity."[12]

The Sustainability City pLAn (pLAn) outlined near-term (2017) and long-term (2025 and 2035) targets across 14 issue areas.[13] The Mayor appointed a Chief Sustainability Officer to oversee the development of the pLAn. To provide a more comprehensive and informed pLAn, the Chief Sustainability Officer created a sustainability team comprising representatives from various city departments as well as private consultants. Each department was consulted during the development of the pLAn to provide insights into their sector, as well as for their continued role and responsibilities in meeting the targets of the plan.

For instance, Los Angeles Department of Water and Power would be leading the effort to develop strategies and programs for conserving and improving the quality of water in Los Angeles. Departments provided important information and data on existing and future projects and programs within their sector, such as the construction of the water reclamation facility designed catch and clean stormwater runoff from Los Angeles City Airport (LAX) in a groundwater basin to replenish the water supply and create a five-acre open park space.[14] The data and information guided the development of the metrics and methodology for the pLAn as well as future reporting.

City staff consulted with community stakeholders, including environmental nonprofits, neighborhood groups, foundations, and businesses to develop the pLAn's

---

[11] Elkington, John, 1994. Toward the Sustainable Corporation: Win-Win-Win Business Strategies for Sustainable Development.
[12] Garcetti, Eric, 2015. Introducing the Sustainable City Plan. https://www.discoverlosangeles.com/travel/the-sustainable-city-plan-of-los-angeles.
[13] Plan: Transforming Los Angeles, 2015. https://www.lamayor.org/mayor-launches-las-first-ever-sustainable-city-plan.
[14] Sustainable City Plan, 2016. First Annual Report 2015—16: http://plan.laymayor.org.

framework. These external stakeholders provided more local information and perspectives on individual neighborhoods and communities. Stakeholders and members of the public were invited to attend workshops and outreach events, where they were able to tell the city's sustainability team what they believed to be the existing challenges to sustainability and where the city should prioritize their efforts. Stakeholders noted a number of different environmental, social, and economic challenges that their communities faced including, but not limited to poor air quality, unemployment, lack of affordable housing, vehicle congestion, lack of green and community spaces, lower services in low-income neighborhoods, and expensive utilities. The sustainability team members collected the feedback from all of these sessions consolidated it and incorporated it into the pLAn.

Based on the input from the departments and the stakeholder feedback, the pLAn focused on issues relating to water, solar, energy efficient buildings, climate, waste, housing and development, mobility, green jobs, resiliency, air quality, environmental justice, urban ecosystems, and livable neighborhoods. Each of the 14 focus areas presented a larger visions statement of what Los Angeles wants to achieve over the next 20 years. Under each vision statement, the plan outlined specific outcomes, which are measurable, quantitative, and time bound deliverables that will enable Los Angeles to reach its overall sustainable vision. An example of an outcome from the water section is for Los Angeles to reduce average per capita potable water use by 20% in 2017.[15] The sections also included success stories or programs that have been or will be implemented to enable the city to achieve these outcomes, such as the Save the Drop campaign, which incentivized Los Angeles residents to reduce water usage through smart irrigation controls and installing low-flow water fixtures. The Sustainability City pLAn highlighted the connection between a citywide sustainable vision and tangible actions and projects that would enable the city to realize that vision.

Implementation

In addition to creating the Sustainability City pLAn, the Mayor charged the Chief Sustainability Officer along with the sustainability team department representatives with putting the plan into action. They appointed sustainability officers in key departments to report directly to the city's General Manager in collaboration with the Chief Sustainability Officer to help achieve the sustainability outcomes. The city's General Manager was charged with reviewing overall progress and alignment with the pLAn's vision. To ensure accountability the General Manager was required to incorporate pLAn progress and outcomes into the General Manager's annual city performance reviews.

Implementing the plan also helped the city to establish budget priorities for sustainability projects in the Mayor's annual proposed budget and directing the General Manager to submit budget proposals that prioritized funding for near- and long-term

---

[15] Sustainable City Plan, 2016. First Annual Report 2015—16: http://plan.laymayor.org.

pLAn outcomes. In addition to budgetary prioritization, the pLAn provided transparency on its progress with the publication of the metrics and data used in the plan on the city's website.[16] Additionally, the pLAn required an annual report on progress on each outcome and its implementation throughout the city. The annual reporting of progress toward LA's sustainable vision is provided by the city departments including the following:

- Department of Aging
- Department of City Planning
- Bureau of Engineering
- Bureau of Street Lighting
- Department of Animal Services
- Los Angeles Police Department
- Los Angeles Fire Department
- Bureau of Street Services
- Port of Los Angeles
- General Services Department
- Department of Transportation
- Bureau of Contract Administration
- Los Angeles Public Library
- Housing and Community Investment Department
- Department of Water and Power
- Department of Cultural Affairs
- Department of Recreation and Parks
- Information Technology Agency
- Los Angeles World Airports
- Economic and Workforce Development Department
- Bureau of Sanitation
- Department of Neighborhood Empowerment
- Los Angeles Zoo

The annual report includes best practices and lessons learned to adapt the pLAn to the changing needs of the city. In addition to the annual reports, the pLAn is intended to be evaluated and updated every 4 years.

Adopt the pLAn

In addition to apply city staff and resources to meeting the overall sustainability vision, the pLAn also continued to engage external stakeholders to support pLAn implementation. The pLAn invited external stakeholders to become partners and "adopt the pLAn," to support progress on an outcome, initiative, or project related to the pLAn.

---

[16] The Sustainable City Plan online dashboard. https://www.discoverlosangeles.com/travel/the-sustainable-city-plan-of-los-angeles.

In 2015, over 60 nonprofits, businesses, universities, schools, and individuals committed to adopting the pLAn and layout specific strategies or projects in which they will support pLAn outcomes. In October 2015, the Los Angeles chapter of Habitat for Humanity adopted the pLAn and made a commitment to increase and improve sustainable housing in Los Angeles. Habitat for Humanity installed synthetic grass, drought tolerant plants, as well as installed new low-flow toilets and showerheads in over a thousand homes to reduce water usage and lower water bills for residents.[17] Engaging local stakeholders as partners in the pLAn ensured collaboration between the city and the communities they served as well as facilitated broader participation.

---

[17] Sustainable City Plan, 2016. First Annual Report 2015–16. http://plan.laymayor.org.

Global and international policies: UN Paris Accord UN G19 and G20

Woodrow W. Clark II, MA³, PhD

Founder/Managing Director, Clark Strategic Partners, Beverly Hills, CA, United States

Background

In 2019, something important happened. Despite the horrific mass murder of over 129 people in three different areas of Paris, France on Friday, November 13, 2018 and frequent killings of innocent people by some who have access and the use of weapons, the world wants to move ahead to global solutions for these kinds of attacks. The UN had planned a Conference in Paris that is focused on the solutions to climate change despite the current (2018) President of the United States, who is even now getting more involved personally as well as his staff in the UN G20 (from which he withdrew the United States in June 2016) now to both enhance trade but also work with other nations.

Below are the Reports from 2016 to 17 and how much is now going on that reflects all of this despite the United States.

United Nations Group of 20 nations (G20) **Summit in Hangzhou, China, September 3−5, 2016**: where these global leaders pledged support of both the United Nations 2030 Agenda for Sustainable Development and the December 2015 Paris Accord for dealing with greenhouse gases, emissions mitigation, adaptation, and finance.

In April 2016, Sherpas drafted the first Presidency Statement on Climate Change. That statement focused on how the G20 could take the lead in promoting the implementation of the Paris Accord and quickly bring it into force.

The second key to G20 support was advice from the Business 20 (B20) that held meetings in the 2 days before the G20 meetings. The B20 provides a significant platform for the international business community to support the work of the G20 by hosting focused policy discussions and developing recommendations geared toward strong, sustainable, and balanced growth in the global economy.

Then, Mr. Xi Jinping, President of the People's Republic of China, made the keynote speech at the B20 Opening Ceremony on September 2. Titled "A New Starting Point for China's Development, A New Blueprint for Global Growth," President Xi emphasized strongly the need for promoting "green development" to achieve better economic performance while mitigating and adapting to climate change (a concept reflected in China's new Five Year Plan, #13, adopted in March 2016). "Green development" is the Chinese terminology for what the European Union calls the circular economy, whereby products, goods, and services are all seen as being reused and recycled to cut down on transportation, use less fossil fuels, and minimize waste all to reduce greenhouse gas emissions.

The conclusions the B20 forwarded to the G20 were titled "Toward an Innovative, Invigorated, Interconnected and Inclusive World Economy," and are included in the recommendations by the G20 in its Executive Summary.

The B20/G20 meetings resulted in the G20 Action Plan on the 2030 Agenda for Sustainable Development that recognizes the global importance of stopping climate change. From the Action Plan: "G20 efforts will continue."

The United States withdraws under President Trump in June 2017 so that the G20 becomes G19 just as the G20 Report from Germa Bonato, Danielo. ReMedia company (http://www.consorzioremedia.it/) and Ecosystems in the EU (http://www.repic.co.uk/).

The drama continues today (2019) between the biggest economies in the world. The results are not known yet—Stand by.

Predictions
Initiatives that cannot be measured if not started

For the past few years since founding Soofa, we have been part of big and small smart cities pilot projects and have collaborated with nearly 120 cities around the world. Unfortunately, over the years, the term "pilot" has become much less about being the critical, rigorously tested and measured first phase of a broader deployment of smart cities technology, and instead is used to simply make something feel like a risk free way to just get something done. It is a mechanism to get something done quickly but not necessarily make that thing last.

Although this is helpful to accelerate the number of new technology implementations in the short run (i.e., cities have grown their pipeline of innovation projects and built up the top of their innovation funnels), it is not helpful in ensuring the sustainability of those implementations in the long run.

That said, in 2019, we will see pilot projects vetted more rigorously before being deployed, and a clearer goals set up front by diverse stakeholder groups, including numerous city departments, private sector partners, and, most importantly, the public. This will mean fewer projects will be launched overall, but those that are will be more likely to get operationalized and scaled up to have a greater impact for the everyday lives of citizens living in the smart city.

The City of Austin's SmartMobility group has a Public—Private Partnership Opportunity Expression of Interest Form, which will likely emerge in more cities as a way to evaluate pilot projects more objectively and ensure they actually have the potential to solve a real problem facing citizens, at scale.

There will be many more interdepartmental smart city RFIs and RFPs

As the shift toward rigorously measured pilot projects happens in 2019, so will a shift toward more holistically written smart city Request for Proposal (RFPs) and Request for Information (RFIs). We are already seeing this emerge with the latest city and county of Denver Smart City RFP.

For all smart cities projects to be successfully implemented, iterated on, and operationalized across the city in a way that tangibly benefits their citizens, these projects need to go beyond a single department so that their applications and benefits to other entities in the city can be realized.

2019 will be the year when more projects than ever advance beyond an innovation team or smart city coordinator, serving as more than just a press release. As more structured learnings emerge from well-executed pilot projects and smart cities working groups within cities throughout 2019, we will see more cities issuing smart city RFIs and RFPs that are multidepartmental in nature and therefore, address a diverse set of problems and challenges that can be solved by the deployment of a single technology.

2019 will be the year where the top-down and bottom-up approaches to smart city development meet in the middle. What is more, we would not be surprised to see these RFIs and RFPs dropping the "smart city," and instead will be focused on the actual problems emerging or fully commercialized technologies can solve.

Public—private partnerships will diversify and engage the long tail of SMBs

Public and private partnerships will extend to the long tail of Small Medium Business (SMBs) in 2019. Infrastructure like digital outdoor communication platforms and kiosks have risen to prominence ever since the City of New York transformed its existing public pay phones to digital kiosks. The business model, which is to provide public amenities free of charge in exchange for the right to sell advertising, has been replicated in dozens of other cities across the globe. However, for as much as this model currently promotes the fact that it supports small businesses and promotes economic development, we will see a new business model emerge, one that will be funded by small business advertising revenue.

Small and medium sized businesses play a critical role in giving neighborhoods and cities their identities; if given an opportunity that is mutually beneficial and valuable, these businesses can add a new source of funding into dynamic public—private partnerships. This will be led in part by a shift in the initial financing mechanisms of kiosks and public communication platforms and the costs of the devices themselves. Lower cost devices, combined with capital from investors who care deeply about the experience of living in and visiting cities will help create the shift toward engaging the long tail in financing a return on smart city infrastructure.

Cities and vendors will lead with the needs of citizens

2019 is when citizen-facing smart cities efforts—especially Internet of Things (IoT) efforts—will be driven from the bottom-up, not top-down. There was a time, not so long ago, when many smart cities efforts could be legitimately accused of chasing the newest, shiniest things, rather than addressing the most pressing problems. Caught up in the excitement of the early days of a promising new industry, it was too common to pursue technology just for technology's sake, rather than thinking critically about which problems government can and should address with better use of data and technology.

Great progress was made on this front in 2018, but 2019 is the year that this dynamic will change once and for all. Ideally, all smart cities efforts begin with asking residents, in one way or another: what are the most significant problems in your day-to-day life that the city can improve by better use of data and technology? What are the city services you would most like to see improved and why? This phrase, "lead with the need," and the upfront and ongoing, meaningful community engagement it reflects, is absolutely critical to ensuring that smart cities efforts are both inclusive and responsive to residents' needs.

Innovation will go deeper in cities, beyond smart city groups and into all departments

In many larger cities, innovation and technology efforts have been led by Chief Information and Chief Innovation Officers and in smaller cities, by city managers and sometimes mayors. Often times, the departments that are tasked with actually *implementing* initiatives and projects (public works, planning, engineering, etc.) have been hesitant and skeptical of smart cities efforts.

Historically, that skepticism has made sense, given the incredible responsibilities of departments like public works. However, as the outcomes of smart cities efforts are becoming more and more measurable and more and more impressive—and as billion-dollar deferred maintenance backlogs become even more common in US cities—2019 will be the year that these historically traditional departments will take note and begin to seriously consider smart cities as a possible money saver. Metropolitan Sewer District of Greater Cincinnati, for example, has had tremendous early success in using sensors in its water system to reduce combined sewer overflows and is betting that smart technology will be a key to complying with its EPA consent decree.

There will be pushback on smart cities conferences from vendors and cities

2018 saw a massive uptick in the number of smart cities conferences. 2019 appears to be on track for more growth. If you wanted to, you could attend a different smart cities conference, trade show, or workshop practically every week. In 2019, the sheer number of smart cities conferences will mean that city officials—and companies!—

will be more discerning about which conferences to attend by determining the ROI for each conference attended and sponsored.

The uptick in similarly focused smart cities conferences also means that 2019 runs the very real risk of creating a smart cities echo chamber … that the same conversations will occur between the same group of industry leaders, just with different city backdrops. Seeking out deeper engagement from city peers with different perspectives will cause city officials to increasingly turn online to city-only learning networks and to in person deep-dive workshop-style events in 2019.

Artificial Intelligence (AI) will prevail, making actionable the treasure trove of data cities have available

Data have no doubt been the fuel powering the smart city industry. As the smart city vertical has grown, solutions rolling out have often had data collection at their core, ranging from IoT sensors to civic engagement platforms—the main use case was collecting information.

Cities have risen to the challenge, investing resources and brainpower into building amazing data infrastructure. Moreover, now—they are beginning to realize that data alone is not enough. 2019 will be all about making this data actionable, and Artificial Intelligence (AI) is fast becoming the tool of choice. Once the subject of Hollywood sci fi hits, thanks to recent advances in the field, AI has finally matured into accessible and replicable use cases that make sense—ranging from predicting the next car crash (like Waycare) to understanding wide scale resident sentiment like we do at ZenCity.

In 2019, AI-based analysis will become the standard for leveraging the treasure trove of data cities have, by generating actionable, in-depth insights out of vast amounts of information. We will see AI being used on a wide scale for automation, decision-making, and civic engagement. Moreover, that will only be the tip of the iceberg of this exciting world of capabilities.

"Off the shelf" solutions will cater to small and midsize cities

Traditionally, the golden rule of smart city implementations has been that they take a long time and that they are tailor-made. Customized, long-term projects have been the bread and butter of smart city solution providers.

However, 2019 will see a positive shift in a new direction. As the industry is maturing, we will start seeing more and more "off the shelf" solutions—SaaS platforms, end-to-end IoT solutions, mass production hardware, and more, make up for a growing chunk of the industry.

This makes sense—technology is evolving and as best practices are discovered, they can be easily duplicated as cities often share similar challenges. The change provides great opportunities on both the city side and the business side. Business will leverage the effectiveness, replicability, and easy tweaking of SaaS, platforms, and other easily scalable solutions.

The biggest winners from this change will be SMCs (small and medium cities), who in fact make up the vast majority of the cities globally. These SMCs, who often lack the resources and/or know-how to embark on tailor-made projects, will finally be able to easily tap into technological solutions that are cost efficient and that have already been tested by other cities. This will create tremendous benefit, leapfrogging many communities into the smart city space, positively impacting the quality of life for more citizens in more places, and opening new markets for vendors. The wide reach of such solutions will in itself create new *benchmarks* for smart city services— benchmarks cities will strive to adhere to.

Cities and vendors will be more accountable for privacy than ever before

2018 was a landmark year for data privacy. This year we saw the best and the worst of how businesses, governments, and individuals around the world are grappling with privacy—ranging from the Cambridge Analytica scandal on the one hand, to the EU's comprehensive General Data Protection Regulation (GDPR) roll out, on the other. If privacy is a concern in the private sector, the importance of protecting privacy increases tenfold when it comes to government and the public sector, and so do consumer expectations.

2019 will be the year when cities and residents alike will start asking the hard questions about how and why different types of data are being collected and used. As residents hold all levels of governments accountable, so will cities turn to the vendors they work with and demand a higher standard of data privacy. What is the 2019 rule of thumb in the smart city space? We believe it is not to single out an individual unless necessary.

2019 will be the year of replicable, scalable smart city innovations

Overall, we are optimistic that 2019 will see some of the most successful and carefully measured smart city projects launch, grow, and be sustained with the support of citizens. If done right, we would not be surprised to see the term "smart city" fade away, as the focus of cities and private sector companies shifts toward solving real problems for citizens—the smart city should become the contemporary city, one that solves its most pressing problems effectively, efficiently, and sustainably—both environmentally and economically.

Globalism and regionalism: overview

Woodrow W. Clark, II MA[3] PhD

Products:

Starbucks Coffee
Solar—RMI
Apple
Off-grid—Case of Sweden

Improving interconnectivity with multimodal transportation

11

Akima Cornell, PhD

Principal, Akima Consulting, LLC, Los Angeles, CA, United States

Overview

The US Environmental Protection Agency (EPA) states that the largest source of greenhouse gas (GHG) emissions in the United States is burning fossil fuels for electricity, heat, and transportation.[1] Transportation is the largest source of GHG emissions, which accounted for 29% of the GHG emissions of the United States in 2017.

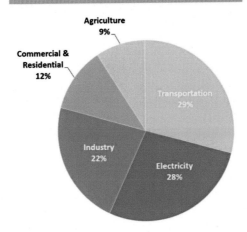

US EPA Greenhouse Gas Emissions.

Source: https://www.epa.gov/ghgemissions/sources-greenhouse-gas-emissions#transportation.

---

[1] US Environmental Protection Agency (2017) Sources of Greenhouse Gas Emissions: https://www.epa.gov/ghgemissions/.

Sustainable Mega City Communities. https://doi.org/10.1016/B978-0-12-818793-7.00011-1

GHG emissions from transportation predominately come from burning gasoline or diesel in vehicles, trucks, ships, trains, and airplanes.[2] As cities move forward with developing and implementing plans to become more sustainable, they must address their largest source of GHG emissions—transportation. This requires developing zero-emission vehicles and infrastructure, increasing public transportation, and improving low-emission or zero emissions transportation options. Cities are seizing on number of opportunities to reduce GHG emissions associated with transportation—from improving fuel efficiency, reducing vehicle use, increasing ridesharing, switching to alternative fuels and vehicles, improving public transportation, and incentivizing active transportation.

Cities are endeavoring to provide people with a variety options to meet their daily transportation needs beyond single-occupancy vehicles. Building on existing relationships with municipal, regional, and local transportation authorities to enhance and expand services will be a critical part of developing sustainable transportation. This chapter discusses how cities are using multimodal transportation and new technologies to meet the changing needs of residents and visitors to improve local and regional transit.

Multimodal transportation

Multimodal transportation refers to planning that integrates various modes of mobility, including walking, cycling, scootering, buses, shuttles, public transit, etc. How are cities developing and incorporating new transportation methods to improve connectivity and reduce emissions? Supporting new multimodal transportation and integrating it into a city requires developing new infrastructure and services. Ideally with the development of new infrastructure and support services for multiple modes of transportation, it will increase the use of the public transportation system and reduce GHG emissions from vehicles.

Developing multimodal transportation requires understanding the travel patterns and transportation needs of the public—where are they traveling to/from, how far are they traveling, what are their transit needs? Understanding how the majority of residents and visitors travel within a city is critical to developing multimodal transit to enhance an existing transit system.

Additionally, cities and transportation planners need to recognize and accommodate the needs of all public transportation users, including families, older adults, children, and people with disabilities. Inclusion and accessibility for all riders strengthen the system and increases ridership. The goal of multimodal transportation is to offer a variety accessible transport services to meet the multitude of transportation needs.

---

[2] Solomon, S., D. Qin, M. Manning, Z. Chen, M. Marquis, K.B. Averyt, M. Tignor and H.L. Miller (eds.) (2007) Fourth Assessment.

Local transportation
First/last mile solutions

Most mega cities have some form of public transportation system—buses, trains, subways/light rail, shuttles, etc. Yet, a fair number of the people in those cities live too far from main transit lines or mobility centers to access them without driving. Similarly, their jobs, schools, shopping centers, or other destinations are too far from transportation to be accessible without a vehicle. This is referred to as the first/last mile problem, wherein individuals have difficulty getting to or from major public transportation routes rendering them more likely to use another form of transportation, like a vehicle, to complete their journey.[3] Furthermore, transit-dependent riders are forced to contend with long walks, circuitously long routes, and extended travel times to complete their journeys.

Planning transit-accessible cities that will reduce GHG emissions will need to integrate a multitude of services that are supportive of transit needs and integrated into neighborhoods around transit stops. Transit access needs to be optimized with multimodal transportation to develop first/last mile solutions. Strategies for designing supportive services and infrastructure within the vicinity of major transit routes and station facilities are key from connecting riders. Partnerships between cities, counties, and other transit agencies are critical to improving connectivity. Through joint cooperation across the public and private sectors, cities can promote a "seamless journey"[4] transit for riders and first/last mile solutions that can compete with convenience of a vehicle. The selection of strategies implemented in any particular city will depend upon the specific transpiration needs of the area with a goal of achieving a high level of customer satisfaction across multiple rider types, including choice riders and transit-dependent riders.

When developing first/last mile solutions, cities need to coordinate with transportation authorities to identify key locations for transit connectivity. Cities must also create cost estimates by strategy and location, implementation timelines, program governance and oversight, fare payment, and reservation technologies. This includes developing new ticketing systems or fares for riders that are connecting to multiple forms of transportation (e.g., using a bikeshare to ride to a train station). Additional actions to optimize transit access and increase multimodal usage include developing infrastructure improvements, changing in transit service levels, and creating public-private partnerships to support multimodal transit options from a wide range of service operators.

[3] Los Angeles County Metropolitan Transportation Authority (2019) "Metro First/Last Mile," website: https://www.metro.net/projects/first-last/.

[4] Transportation Board of the National Academies (2003) Report 94 - Fair Policies, Structures and Technologies: Update, Transit Cooperative Research Program, Federal Transit Administration. Multi-Systems, Inc., Cambridge, MA.

Los Angeles Metro Bikeshare Program.

It is also important to recognize that competing demands for funding and physical space within a city will often require complex trade-offs. Developing first/last mile strategies must take into account cost-effectiveness, station context, and an access hierarchy that prioritizes alternative modes of transit. Cities may need to seek out potential funding sources for infrastructure improvements, including federal, state, and local funds as well as opportunities for private sector investments.

Regional connectivity

Growing US cities has led to urban sprawl, and a larger dispersal of travel, cities, and transit agencies have faced numerous challenges in providing increased service and access across wider regions.[5] Many cities across the country have multiple transit agencies and companies providing transportation services, each limited by their service boundaries. Yet, riders often need to use multiple systems to reach their destinations. Often riders are reliant on one or more transit operators with a disconnected fare system, which means longer commutes and higher costs.

City and transportation planners need to focus on developing multimodal transportation with partnerships between multiple transit agencies and private businesses in order to provide seamless travel for riders traveling across regions.[6] Developing

---

[5] Transportation Board of the National Academies (2003) Report 94 - Fair Policies, Structures and Technologies: Update, Transit Cooperative Research Program, Federal Transit Administration. Multi Systems, Inc., Cambridge, MA.

[6] Transportation Board of the National Academies (2003) Report 94 - Fair Policies, Structures and Technologies: Update," Transit Cooperative Research Program, Federal Transit Administration. Multi-Systems, Inc., Cambridge, MA.

connectivity across a region requires establishing partnerships and operating agreements among cities and transit operators. Cities, transportation authorities, and private transit organizations will need to coordinate and agree on the fundamental aspects of developing and operating a regional transit system, which includes the following[7]:

- Policy and partnerships—Which agencies participate in the regional system? How is it operated? Who is responsible for maintaining the system?
- Ridership—How will riders use the system? What are the benefits and challenges?
- Ticketing—What types of tickets and fares will be available and are they compatible throughout the system?
- Costs and revenue—What are the initial and operational costs associated with developing a regional system? Which agency is going to pay for upgrades and improvements? How will revenue and resources be allocated among regional partners?

To integrate fares and transit services across a region with multiple agencies, partnership agreements are needed to establish responsibilities, ownership, and allocation of costs and revenues. Cities may seek to create an independent regional commission consisting of representatives from all transit agencies and key stakeholders to resolve disputes and coordinate policies, fares, services, and technologies. Regional commissions are able to create coordinated decision-making among multiple agencies ensuring consistency of services and improvements as new modes of transportation and technologies become available.

---

[7] Joslin, Ann (2010) Regional Fare Policy and Fare Allocation, Innovations in Fare Equipment and Data Collection, National Center for Transit Research. University of South Florida, March 2010.

Smart cards and mobile apps

Choosing a fare technology or ticketing system is a key aspect to integrating multiple transit services and implementing regional transit.[8] Oftentimes existing fare technology and ticketing systems need to be upgraded in order to accommodate a more complex multimodal regional transit system. Determining which technology is appropriate for each system depends on existing fares and ticketing options. Many transit agencies have magnetic stripe paper cards that offer a specific fare (stored value, 1-day pass, etc.), but these tickets are limited and cannot be used with multimodal transportation that requires different fares.

Hop Fastpass.

Source: https://myhopcard.com/home/.

Smart cards or mobile ticketing are better suited for a regional system with a variety fare structures (e.g., single-ride, 1-day pass, monthly pass, etc.) and multimodal services. A smart card is a reusable transit ticket that can store a multitude of data,

---

[8] Shah, Vivek, Heiligman, Rachel, and Miller, Alexandra (2015) Achieving Regional Fare Integration in New Orleans: Innovative Cost Sharing Arrangements and Technologies, University of New Orleans Transportation Institute (UNOTI) Department of Planning and Urban Studies.

such as ticket purchases and passes that can be read by different ticket verification machines within a regional transit system. Utilizing a single fare media or smart card, which incorporates all fare and transit services, provides ease of use and a better rider experience. The Hop Fastpass used in Portland by TriMet, C-TRAN, and Portland Streetcar is an example of a regional smart cards that enable riders to transition between transit services run by different transit agencies.[9] Riders can travel across regional boarders and various transit systems with a single smart card. Another advantage of utilizing a smart card is the ability to access multimodal transportation options on a single card. The Los Angeles County Metropolitan Transportation Authority (Metro) Transit Access Pass (TAP) card enables riders to multiple transportation services including bikeshare, buses, and rail.[10]

A number of transportation agencies are developing mobile phone applications (mobile apps) to enable customers to purchase and validate transit passes on their smartphones. Mobile apps can provide riders with a variety of ticket options. Once a customer downloads the app and inputs their payment information, they can purchase any ticket and activate it upon boarding. The electronic ticket is validated upon entry and verified by the driver or conductor. Mobile app enables transit agencies to include regional passes connecting multiple transit agencies as a ticket option. Additionally, mobile apps can offer multimodal transportation services connected through one mobile app, making it easier for riders to transfer between different transportation services. Additional benefits associated with offering a regional pass and multimodal transportation on a mobile app are the following:

- It enables customers to buy tickets directly on their phones—anywhere, anytime, without having to go to a ticket office.
- Operations and maintenance on a mobile app is relatively low and conducted by the mobile app developer.
- Transit agencies do not need to maintain ticket stock and printers.
- Customers can purchase tickets in advance.
- Drivers do not need to provide receipts or change on cash transactions.
- Increased marketing and branding opportunities to encourage residents and visitors to use transit.

Utilizing smart cards and mobile apps reduces the barrier for riders to use multimodal and regional transportation by making it easier to purchase tickets and transfer between transit services. Riders can purchase tickets in advance and better plan their routes providing for a seamless journey. Cities and transportation agencies can also offer discount fares or new ticket options to better meet the transit needs of their riders.

---

[9] Portland, Oregon, Hop Fastpass: https://myhopcard.com/home/.
[10] Los Angeles County Metropolitan Transportation Authority, TAP Card: https://www.taptogo.net/.

Case Studies

New Orleans

In the decades, post-Katrina, people and businesses have been returning to the Jefferson and New Orleans parishes. According to the US Census, approximately 53,000 New Orleans parish residents commute to work daily, and approximately 87,000 residents from outside the parish commute into New Orleans for work.[11] As a result, there was a renewed focus on regional transit coordination. Changes in ridership demand for New Orleans Regional Transit Authority (NORTA) and Jefferson Parish Transit's (JeT) need to be reevaluated to reflect increasing demand for regional transit.

NORTA and the Greater New Orleans region have some history in regional transit coordination. Prior to 2005, NORTA and JeT offered a regional day pass allowing for trips across both services. The pass was in place between 1999 and

---

[11] US Census Bureau Center for Economic Studies, 2012.

2004, when it was discontinued through mutual agreement by the agencies. NORTA and JeT also engage in regional service planning, attempting to align connecting bus schedules. As of 2016, NORTA provides service along 34 bus lines, five streetcar lines, paratransit services, and is supporting operations of two ferry services. JeT has 12 bus routes and paratransit services.

In 2012, the Regional Planning Commission (RPC) commissioned a Comprehensive Operational Analysis (COA) to collect ridership data on NORTA and JeT services. The 2012 survey asked riders about their travel behavior, transfer behavior, origin address, destination address, and demographic information. In total, the survey contains responses from 7225 riders. The survey results showed that typical riders are transit-dependent ethnic minorities of low income. The survey also noted that majority of existing transit patrons transfer at least once to reach their destination. Survey responses from regional stakeholders also identified regional transit as key for creating a seamless transit experience for riders. Stakeholders supported updating regional transit services and routes to meet the changes in neighborhood demographics, travel patterns, and new economic centers.

NORTA's services include network of buses and streetcars, as well as ferry crossings throughout New Orleans Parish with a few routes into Jefferson Parish. NORTA offers one-way fares and passes that allow riders to purchase tickets to meet their travel needs. Jazzy Passes allow the rider to purchase a fare ahead of time and are inactive until it is activated onboard a bus or streetcar. Jazzy Passes remain valid for the specified number of days after the pass is first activated. A 1-day Jazzy Pass is valid for 24 hours after it is first activated. For frequent riders, purchasing a Jazzy Pass improves convenience and reduces the overall cost of travel. Yet, these passes were not transferable to the JET transit services, which required purchasing a separate ticket.

In July 2018, NORTA launched an updated version of the GoMobile app, which enables riders to track buses and streetcars in real time. The new app locates buses and streetcars with an automated vehicle location system, which was installed on each of the fixed-route buses and streetcars. In addition to installing the tracking device, each bus and streetcar was outfitted with an electronic ticket scanner. The app tracks buses on a map that shows their real-time location within 30 minutes or less from a rider's location.

The app also enables riders to buy and store electronic bus, streetcar, and ferry tickets for up to a year. Purchased mobile tickets can be used with the new electronic scanners found on each bus, streetcar, and ferry terminals. The GoMobile app also offers a regional transit bus ticket, which is an all-day pass that allows riders to transfer between JET and NORTA transit services, enabling riders to transfer between the two service operators, improving regional transit.

City of Los Angeles

In March 2010, the cities of Los Angeles and Long Beach collaborated with the Los Angeles County Metropolitan Transportation Authority (Metro) and the City of Los

Angeles Department of Transportation to develop a new first/last mile multimodal transportation options system at designated transit station locations. The goal was to provide transportation options to residents of urbanized and nonurbanized areas regardless of income. Additionally, the system would enhance the access of low-income individuals and welfare recipients to employment centers, job training sites, community colleges, and other educational facilities. The new first/last mile multi-modal transit system integrated a suite of transportation services, including secured bicycle parking, bikesharing, shuttles connecting riders directly to existing bus and rail services.

Metro rail stations have the highest concentration of riders and the largest need for first/last mile connectivity. Metro rail stations have high ridership, which could be enhanced with multimodal transportation services. Metro rail station and services could include all or a combination of the following: bicycle parking, bikesharing, ridesharing, scooter, and shuttle services. Multimodal services could be accessed through smartphone technology or account-based card. Such services include the following:

- Secured bicycle parking—Enable rider to use their own bikes to get to and from stations and improve the multimodal infrastructure at select Metro rail stations through the provision of bike parking facilities, which could be within an enclosed space that is free standing, built-in, or at a storefront.
- Bikesharing—Bikesharing services provided through the Metro Countywide Bikeshare Program at Metro rail and bus main transit stations and throughout services areas. Bikesharing is a program designed for point-to-point local trips using a shared use fleet of bicycles that are strategically located at docking stations throughout a well-defined area and within easy access to each other. Bikesharing services in the City of Los Angeles will be provided through Metro's Countywide Bikeshare Program, while bikesharing services in the City of Long Beach will be provided through Long Beach's Bikesharing Program.
- Carsharing—Select parking spots at or near the select stations will be rented out to carsharing companies for operations. With carsharing, travelers can use a car for a specified amount of time and then return the car to a designated carsharing location.[12]
- Ridesharing—Dynamic ridesharing companies have the capability to provide real-time first/last mile connections, offering areas for ridesharing pick-up/drop offer services adjacent to Metro rail and bus stations.

---

[12] Currently, Metro has an agreement with a carsharing company to make 20 cars available at each of the 10 locations. Carsharing will be located near selected Metro stops along the Blue, Expo, Gold, Green, Orange, and Red Metro lines. Additional information can be found at http://thesource.metro.net/2015/05/29/zipcar-adds-more-car-sharing-locations-near-metro-station/.

TAP card: https://www.metrolinktrains.com/ticketsOverview/where-to-buy/by-mail/tap-transit-access-pass/

Over time Metro has coordinated with other regional transit agencies throughout Southern California to make the TAP card the primary ticketing option. Riders can easily transfer between several transportation services creating a larger regional system. As of 2015, riders can use their TAP card on the following services[13]:

- Access
- Antelope Valley Transit Authority
- Baldwin Park Trust
- Beach Cities Transit
- Burkbank Bus
- Carson Circuit
- City of Monterey Park Spirit Bus
- Compton Renaissance Transit Systems
- Culver CityBus
- Foothill Transit
- Glendale Beeline
- GTrans (Gardena)
- Huntington Park Transit Unlimited
- LA County Department of Public Works
- Long Beach Transit

---

[13] Go Metro with TAP: https://www.metro.net/riding/fares/tap/.

- Los Angeles Department of Transportation
- Los Angeles World Airports
- Metro
- Metrolink
- Montebello Bus Lines
- Norwalk Transit Systems
- Palos Verdes Peninsula Transit Authority
- Pasadena Transit
- Santa Clarita Transit
- Santa Monica Big Blue Bus
- Torrance Transit

Riders load stored value (i.e., cash amount) or a multitude of day, week, or month passes on their TAP cards using the ticket vending machines and can transfer between agencies and multimodal services. Transferring fares are automatically deducted by the second transit agency when riders "TAP again" on the ticket validator. An integrated TAP card benefits riders by using a single card throughout the system and not having to carry multiple tickets or exact change.

In 2018, Metro announced the integration of the regional bikeshare program with Metro's TAP smart card program.[14] The motivation for the integration was that many users of bikeshare programs in Los Angeles County also use Metro's bus and rail services. Bikeshare will provide additional first/last mile services to help travelers connect with their final origins and destinations. Metro has worked to develop strategies to enhance TAP card functionality, and reviewing vendor technical specifications for those enhancements.

Integration between the Regional Bike Share program and the TAP program would make bikeshare services easier to use, provide a more seamless user experience from a payment perspective, and expand the reach of bikeshare to additional participants—particularly low income, transit-dependent populations. In addition, integration would provide Metro with additional travel data that could be particularly helpful for transit planning purposes, such as origin and destination information for linked transit and bicycle trips.

Additionally, Metro is adapting the TAP card system to mobility services outside of public transit, including hailing rideshares, space reservations for park and ride, as well as scooter services. Metro is also working on a TAP mobile app, enabling riders to create a single account to pay for rides on trains, buses, bikeshare, among other services as well as incorporate additional payment options.

---

[14] Metro TAP to Go: https://www.taptogo.net/MetroBikeShare.

Leading historical cities

12

Lucia Elsa Maffei[1], Elisa Castoro[2]

[1]*Private Law Practice, Lawyer qualified to the higher Courts, Matera, Italy;* [2]*Professor, Graduated in Foreign Languages and Literatures, English Translator and Photographer, Matera, Italy*

A recognition that aims to strengthen cooperation between cultural operators, artists and European cities or to bring out the richness of cultural diversity in Europe or even to emphasize the common aspects of European cultures.

Matera and its territory, therefore, is called to give answers to the challenges Europe faces, through culture: "Smart growth (integrated urban planning of a city of culture and knowledge that integrates economic development, culture, creativity and digital technologies), sustainable growth (combining creativity and technologies for production and energy efficiency or the exploitation of environmental resources scarcity) and inclusive growth promotion of intercultural dialogue also in light of the changes taking place in the Mediterranean basin, and social inclusion" (http://www.matera-basilicata2019.it/it/mt2019/percorso/come-si-diventa-capitale.html—last consultation February 04, 2018)

The weekly insert cover "La LETTURA" # 319—Corriere della Sera, on newsstands until January 13th, 2018, shows a photograph by Armin Linke (Milano, 1966), portraying

A cloud of vapor from the highest mountain of Java, the erupting Semeru volcano ….. and shows the author's research on the conditions of climate, territories and oceans in a world in which the choices of international policies transform the planet. A powerful image, which will be on display in Matera, for the Anthropocene Observatory exhibition, as part of the program of Matera European Capital of Culture 2019

(LA LETTURA #319—Corriere della Sera—on newsstands until January 13th)

Overview

The history of Matera is complex and fascinating with its being a fusion of landscapes, buildings, civilizations, and cultures. From the rock civilization to those of Byzantine and Eastern origin, to the advent of the Normans, the systematic attempt to reduce the rock city to the rules of the culture of the European city.

The last eight centuries of construction and finishing of the city have tried to shape, to overcome the natural resistance of the preexisting rock habitat, determining architectures and urban arrangements of particular quality and originality, from the Romanesque to the Renaissance, to the Baroque. Today, again under the banner of European urban culture, Matera faces the aspects of the challenge of renewal, sustainable recovery, and regaining of lost identity.

The city lies in a small canyon carved out by the Gravina—a rocky ravine created by a river that is now a small stream, dividing the territory into two areas, the Sassi (the ancient town) and the natural caves and park.

As this photo below shows, Matera was one of the first human settlements in mountains and rock that date back to the Paleolithic. Its architecture is magnificent and unique, consisting of the immediately visible habitations dug into the calcareous rock and the successive stratifications of houses, courtyards, ballotas, palaces, churches, and gardens, and of internal and invisible cisterns, neviere, caves, tunnels, and water control systems. The largest cistern has been found under Piazza Vittorio Veneto.

(Parco della Murgia Materana)—Located in the mountains north of the Mediterranean Sea and near the Italian heel, Matera is inland on the side of a mountain.

Matera preserves a large and diverse collection of buildings related to the Christian faith, including a large number of Rupestrian churches and monasteries, and some are complex cave networks with large underground chambers. Probably, the Rupestrian churches were already places of worship in the rock civilizations that preceded the Christian one.

Courtesy Artist Salvatore Arancio, Motherless Child, Performance per Matera Alberga Project (Matera2019) curated by Francesco Cascino.

View of Matera at night with people celebrating a holiday.

Courtesy Artist Dario Carmentano, La Fonte del Tempo, Permanent artwork per Matera Alberga Project (Matera2019) curated by Francesco Cascino.

Below is a picture of the Cave Church Madonna de Idris whereby the Church is built both on and in the mountain site. This occurred hundreds of years ago. Moreover, the entire city has homes, offices, and businesses built into it. That is what Matera is famous for, as people do live and work in the mountain side today. Visitors from around the world come to see and experience this city and the surrounding areas as today the same historical values and their results exist in Matera. The earth is being saved, lived in, and protected for future generations.

Consider the **Matera Alberga—Hospitable Art,** which is a project for Matera European Capital of Culture 2019 produced by Matera—Basilicata 2019 Foundation, conceived and curated by **Francesco Cascino**, Art Consultant, Founder, and Artistic Director of Arteprima, in collaboration with **Christian Caliandro** and in partnership with Consortium of Matera Hotels, and with the support of the Carical Foundation.

All the installations are permanent and have a distinct participatory component inviting visitors to directly experience the sense of living together for thousands year in harmony. They are designed as permanent projects within the spaces of the hotels involved. In fact, with **Matera 2019**, six hotels in the city of Matera, which evoke the old Neighbourhoods of the Sassi, will become places of hospitality and creative experience, spaces for cultural production, home to encounters between residents and travelers and to social and cultural sharing.

Corte San Pietro—IDRA—Artwork by Alfredo Pirri (external) above and below is Locanda San Martino—Rapporti—Artwork by Filippo Riniolo (Int. Sculpture and Voice).

IDRA, Permanent artwork per Matera Alberga Project (Matera2019) curated by Francesco Cascino.

Reproduced with permission from Michelangelo Camardo.

The majestic Matera Cathedral—dedicated to Santa Maria della Bruna since 1389, the church of St John the Baptist, the Church of S. Domenico, and the Church of Santa Maria della Valle Verde on the Via Appia were built during the Middle Ages. From this moment onwards a real urban nucleus takes shape, initially concentrated around the Cathedral, built in the Apulian Romanesque architectural style (the interior is on the Latin cross plan, with a nave and two aisles The decoration is mainly from the 18th century Baroque restoration, and a recently discovered "Last Judgment" portrait in a Byzantine-style 14th-century fresco) and located at the top of the Civita hill dividing the Sassi in two: the Sasso Barisano facing East and the Sasso Caveoso facing South. Both in the Sasso Caveoso and the Sasso Barisano, there are two other important churches dedicated to the Apostle Peter. San Pietro Barisano is the largest rock church of Matera.

The Church of Santa Maria de Idris is located within the rocky spur of Monterrone overlooking the Sasso Caveoso, near the Church of San Pietro Caveoso and the homonymous square, offering a unique view of the city and the Gravina.

The church of Santa Lucia alle Malve—the first female monastic settlement of the Benedictine order—is located near Santa Maria de Idris in the Malve district. It is one of the most important Rupestrian churches in which there are some of the most beautiful and important wall paintings in the Matera area.

The historic center of Matera is located on a plain bordering the Sassi. Here showing different levels of overlap urban strata. Located below the central square Piazza Vittorio Veneto, there are some openings that show the original level of the places, called hypogea. Dividing into continuity, the hypogea form a real submerged city connected with the Sassi. Here, there are exceptional rock structures such as the large cistern called Palombaro Lungo until recently navigable; and the rock church of the Holy Spirit dating back to the 10th century AD. It is located just beyond the former convent of S. Lucia alla Fontana that divides the 18th-century church of the same name on the right and the Ferdinandea fountain on the left.

The historic centre of Matero.

Moreover, the historic center of Matera is developed with streets connecting several squares along the 18th-century axis. In the center of Matera, there are a series of important palaces and churches such as the church of St. Francis of Assisi—hypogea is also present here under the level of the homonym square. The most important noble and evocative buildings—Palazzo Gattini, Palazzo Venusio, Palazzo Malvinni Malvezzi—unraveling from Piazza del Sedile through the narrow Via Duomo come to surround the Cathedral of Matera which has a Romanesque-Pugliese style but characterized by unique artistic elements—dating back to 1270. The majestic bell tower dominates the whole amazing landscape. Below are mountain rock holes where people used to live.

Sassi di Matera: ancient cave dwellings.

Inside the areas pictured above are rooms, living spaces, and areas like homes today.

Courtesy Artist Georgina Starr, The eternal ear, Permanent artwork per Matera Alberga Project (Matera2019)
curated by Francesco Cascino.

Sextantio—The Eternal Ear—Artwork by Georgina Starr (ritual and voice).

Last but not least, Palazzo Lanfranchi, the National Museum of Medieval and Modern Art of Basilicata, nowadays. Here, follows a series of exhibitions and cultural events of international importance related to culture and history of the territory, but not only.

Case in point: Yuneshima island in resort Osaka city

By Andrew DeWit PhD. dewit@rikkyo.ne.jp.

Overview

Japan is the planned smart resort city on Osaka's Yumeshima island. A group of investors came over to Ikebukuro on the fifth to hear about smart cities and how to integrate the project with what the Japanese are already doing. So I frantically wrote up a summary of the past 2 months' reports and budgets on CHP and SDGs (new report on 1/31), "Super Cities" (new special committee report out on 2/14), 2025 World's Fair and Society 5.0, National Resilience and massive tax breaks/subsidies for smart gear, and loads of other stuff (of course, H2's in there as well).

The summary's the attached ppt. I had to eliminate the name of the investors, as we agreed on nondisclosure.

Osaka is desperate to use their artificial islands as an IoT/AI showcase and growth pole. Moreover, the Yumeshima project proposal is a bracing 1 million square meters (60 ha), so I dug through things I saw years ago about the already installed above ground/underground infrastructure on Yumeshima.

Among other things, I suggested that investors work with Kashiwagi's Cogeneration Centre (which did the CHP and SDGs report), and tweak their proposal and maximize its smartness, drawing on ambient heat energy as well as other Renewable Energy (RE).

Moreover, in the English-Language clip below, you can see where the integrated resort in planned, plus the already installed 10 MW solar farm (some dated details here, in Eng: http://www.kansai.meti.go.jp/2kokusai/Hikarinomori_Power_Plant_English.pdf).

The integrated resort offers are great potential to integrate the variable power of the solar array, whose JPY42/kWh revenue stream is guaranteed until 2033.

Below that clip is a screenshot from the Yumeshima rail station Osaka Metro wants to build (with the resort owners picking up a large share). Pretty impressive, no?

The total project costs are massive, with the resort alone estimated at roughly JPY 930 billion. However, there are now some huge incentives for RE and smart gear, so that is what I added up and inserted in a model of the island. What is especially cool is that the project is part of the 2025 Expo, whose core theme is SDGs. So, smart gear is central, as opposed to merely an add-on.

The annual visits to the resort are assessed at 15 million (to a hotel of 3000 rooms conference hall for 6000, casino, commercial facilities, etc.). The required infrastructure represents a lot of potential heat energy. Same with container terminal and the Expo site.

Different from failed smart cities like Masdar and Songdo, this project has immense potential to serve as the anchor for a revival of Osaka and its region. Sure, a casino is at the heart of the model, but that means people will come (cf Masdar, Songdo, and other stalled projects).

As for gambling per se, a centralized casino would likely be far less energy-intensive than the aggregate of Japan's myriad energy-hog pachinko halls, still a JPY 20 trillion business. Moreover, surely it would save a lot of lives by dramatic cuts in the immense clouds of tobacco smoke (seriously).

References

Matera cited http://www.basilicatanet.com/ita/web/index.asp?nav=matera.
La Lettura. #319 — Corriere Della Sera. January 2018.
https://it.wikipedia.org/wiki/Storia_di_Matera.
http://www.matera-basilicata2019.it/it/mt2019/percorso/come-si-diventa-capitale.html.
Francesco Cascino — Curator.
www.francescocascino.com.
www.arteprima.org.

Tokyo sustainable megacity: robust governance to maximize synergies

Andrew DeWit, PhD

Professor, Kikkyo University, Toshima City, Tokyo, Japan

Introduction

Japan's megacity of Tokyo remains the world's largest. It emphasizes integrated and inclusive governance, which is perhaps its most important lesson to other city regions. Close collaboration allows Tokyo to deploy decarbonizing critical infrastructure networks that simultaneously mitigate and adapt to climate change and resource crises. Tokyo's policymakers work in a multilevel collaboration, which includes all levels of government, and explicitly recognizes that Tokyo and the national community confront increasingly serious crises. Tokyo's planning, policy implementation, and international engagement also show that they are keenly aware of a rapidly urbanizing world whose 20th century energy- and resource-intensive growth paradigm is patently unsustainable. These domestic and international factors give Tokyo (and Japan as a whole) multiple incentives to excel in the development and diffusion of the elements of sustainable megacities. Together with the national government's National Resilience project, Tokyo also stresses global sustainable-development goals and other means of maximizing equitable resource efficiency, to foster industries consistent with global needs.

Tokyo metropolitan government

As of March 31, 2019, the total population of the region directly under the authority of the Tokyo Metropolitan Government (TMG) was 13.86 million. This number represented an increase of over 103,000 compared to the previous year, even as Japan's total population declined by 330,587 to 126.85 million. TMG's spatial scale is 2191 square kilometers, and its economic output (2014) of JPY 94.9 trillion was 19.4% of Japan's GDP. The Tokyo city region, which includes TMG in the larger conurbation of the Kanto region (including Saitama and other neighboring prefectures), remains the world's largest megacity. In 2018, the total population of the Tokyo conurbation, or greater Tokyo, exceeded 37 million. This aggregate population was much larger than second-ranking Delhi, at 28.5 million. Hence, managing Tokyo—whether as

TMG or the larger conurbation—is necessarily a matter of understanding and coping with scale.

In Japan, governing to exploit the opportunities of scale is done in a collaborative, multilevel manner. TMG itself is a regional government that comprises 23 special wards (e.g., Shinjuku-ku, Toshima-ku), 26 cities, 5 towns, and 8 villages. This structure dates back to 1943, at the height of the Pacific War, when the former Tokyo Prefecture and Tokyo City were merged for greater efficiency. TMG's relationship with its wards is comparatively unique, even within Japan. TMG undertakes several of the functions that the wards would ordinarily perform (since they are administratively equivalent to cities), including water supply, fire-fighting, and sewerage services. The metropolitan-wide provision of these services is unusual, because cities ordinarily perform them as part of the urban package of spatially defined public services. It is generally more efficient and effective to leave these functions to the municipal authority, which has the best understanding of local conditions coupled with the closest relationship to citizen voters. However, Tokyo's density of over 15,000 inhabitants/square kilometer within the 600 square kilometer area of the 23 special wards changes the fiscal and administrative math. This density allows service-delivery efficiencies, and intercity equity, to be maximized through a regional system that is simultaneously closely consultative. This collaboration is institutionalized in a Metropolitan-Ward Council.[1]

Robust and integrated governance is on par with advanced technology, as the latter's merits cannot be maximized without the former. We see this fact in TMG's water supply. The water-related systems (e.g., sewerage) in TMG's network are particularly energy intensive and crucial to public health, as they are in any megacity. Moving around huge volumes of water (for TMG, an average of 4.13 million cubic liters per day in 2018), whether for consumption or for waste treatment, requires prodigious amounts of energy. Moreover, poorly designed or badly managed water systems pose enormous health and disaster risks, in addition to wasting precious resources. TMG's replacement of its leaky lead and iron pipe network with stainless steel brought its water loss down from 17% in the 1970s to an incredibly low 2% in the 2010s. TMG's rate of water loss is well under the double-digit leakage rates that OECD reports on "Water and Cities" indicate are common even in developed countries.[2] TMG's capacity to plan and operate water its networks on a regional scale, with institutionalized local consultation, allowed expert information and other scarce resources to be focused on the macrolevel management challenge. Rather than leaving each city to—as it were—reinvent the wheel on its own, strong metropolitan governance afforded the capacity to determine and deploy a best-practice solution.

---

[1] On this point, see p. 153 "The Structure of the Tokyo Metropolitan Government (TMG)," Tokyo Metropolitan Government, 2018: http://www.metro.tokyo.jp/ENGLISII/ABOUT/STRUCTURE/index.htm.

[2] On these items, see the OECD's work on Water and Cities: Ensuring Sustainable Futures: http://www.oecd.org/water/water-and-cities.htm.

The importance of this integration of systems, under the aegis of collaborative governance, needs to be underscored, in light of global sustainability challenges. Experts on urban systems point out that "a 'one-water approach', which seeks to integrate various water supply, treatment and management infrastructures into a single infrastructure system perspective that considers the full life cycle of water provisioning in urban areas, can simultaneously deliver liveability, resilience and sustainability benefits."[3] Scaling up such cross-sector integration is beyond the competency of markets and civil society, whose planning capacity and incentives are woefully inadequate.

Both the market and civil society lack the incentives and capacity to lead. Businesses are not trusted to manage the community and its increasingly data-rich network infrastructure. Moreover, businesses certainly do not have the incentives to overlook their own bottom line and maximize the community's sustainability and resilience in the face of multiple existential threats, particularly climate change and energy insecurity. At the same time, most of civil society is not well versed in the role of critical infrastructures and the threats posed by climate change. Nor is civil society seemed sufficiently organized or financed to do much more than veto initiatives through demonstrations or noncompliance.

Even public governance is not up to the task if it is too fragmented to reach regional, multilevel agreement on financing and other essential issues. However, TMG and Japan's capacity to plan for sustainability has grown increasingly robust. One major development was the 2005 legislative change in the structure and objectives of the National Spatial Plan (NSS). Organizational restructuring and other factors led to a more inclusive planning process that shifted its focus from growth to sustainability. Policymakers explicitly recognized growing resource constraints, and sought to address those. The changes thus deliberately incorporated environmental issues into planning, and created autonomous regional planning. TMG plays a powerful coordinating role, as the regional plans are under the authority of the major subnational governments.[4]

TMG and the national government are also coordinated an expanding portfolio of national and subnational "National Resilience" plans (NRPs) that have legal precedence over other plans.[5] These plans are all-hazard, emphasizing cost-effective cross-sectoral adaptation to multiple risks while also achieving broader socioeconomic sustainability. They are key to realizing the mitigation and adaptation synergies outlined in Fig. 13.1. By May 2019, all of Japan's 47 prefectural

---

[3] See p. 109 "The Weight of Cities: Resource Requirements of Future Urbanization," International Resource Panel, 2018, available at the following URL: http://www.resourcepanel.org/reports/weight-cities.

[4] On this, see (in Japanese) Yada Toshifumi "The National Spatial Planning System's Objectives and Issues," Economic Geography Vol 62, 2016: https://www.jstage.jst.go.jp/article/jaeg/62/4/62_360/_pdf.

[5] The central government's National Resilience plans for 2014—19 are available (in Japanese) here: https://www.cas.go.jp/jp/seisaku/kokudo_kyoujinka/kihon.html.

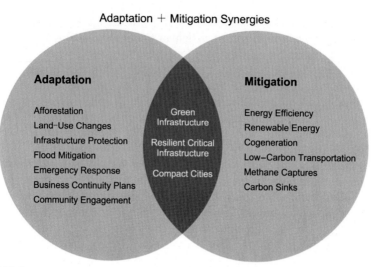

Adaptation + Mitigation Synergies

Adaptation

Afforestation
Land–Use Changes
Infrastructure Protection
Flood Mitigation
Emergency Response
Business Continuity Plans
Community Engagement

Green
Infrastructure

Resilient Critical
Infrastructure

Compact Cities

Mitigation

Energy Efficiency
Renewable Energy
Cogeneration
Low–Carbon Transportation
Methane Captures
Carbon Sinks

FIGURE 13.1

Synergies for sustainability.

Source: Author Andrew DeWit.

authorities, along with 192 cities and towns, had either adopted local versions of the NRP or were formulating plans.[6] As of April 26, 2019, the Japanese national government (through Liberal Democratic Party Secretary General Nikai Toshihiro) has also proposed that this disaster-resilience paradigm become a focus of Japanese-Chinese cooperation on the Belt and Road initiative.[7]

Good governance versus virtue signaling

TMG's balance of broadly inclusive but large-scale governance is critical to achieving sustainability. This assertion may seem so obvious as not to require statement. Yet governance routinely gets short shrift in the emphasis on eye-catching renewable-energy goals, GHG-reduction targets, and other metrics. The sustainability literature is also replete with case studies of microlevel technological and other solutions that fail to ask the important question of whether they can be scaled-up; and if so, precisely how in organizational terms, so as to maximize equity, resource-efficiency, and other essential public goods. Overlooking governance has, if anything, worsened in recent

---

[6] Links to Japan's subnational National Resilience plans are available (in Japanese) here: https://www.cas.go.jp/jp/seisaku/kokudo_kyoujinka/tiiki.html.
[7] See (in Japanese) "Nikai speaks on the Belt and Road initiative," Nihon Keizai Shinbun, April 26, 2019: https://www.nikkei.com/article/DGXMZO44263190W9A420C1EA3000/.

years, even as the multiplicity of climate, equity, and other challenges has increased and their interaction worked to exacerbate one another.

One example of the above were the well-meaning but patently unrealistic bullet points in a megacity communique released in Tokyo in May 2019. Particularly, since the United States Trump Administration's 2016 declaration that it would withdraw from the Paris Agreement, there has been an emphasis on commitments to honor the Agreement by local governments, academic institutions, businesses, indigenous groups, investors, and others.[8] This defiance in the face of dysfunctional national government is understandable. Perhaps, as is often suggested, it is even essential for maintaining momentum and "hope" on climate goals. However, at the same time, it appears to have amplified a tendency—especially in Anglo-America—to compete on virtue signaling versus credible planning. This phenomenon is evident in increasingly ambitious energy and emissions targets. These targets generally omit coordinated planning and overlook other essential sustainability targets, such as adapting health, infrastructure, and other essential urban systems to unavoidable climate impacts. Thus, the U20 Group of Cities (composed largely of megacities), which met in Tokyo on May 22, 2019, committed itself to 100% renewable power by 2030% and 100% renewable energy (i.e., in transport and other nonpower sectors) by 2050.[9]

These renewable energy targets are not persuasive. This is not to deny that renewable energy has an important role in helping to decarbonize megacities as well as entire economies. However, for many reasons (such as lack of space), megacities generally have comparatively low potential as sites for renewable power and energy. Even the German capital of Berlin, the center of the "Energiewende," energy transition, derived only 4% of its power from renewables in 2015, the bulk of that being biomass (the most recent data in 2019).[10] We should add here that—in sharp contrast to TMG and Japan—Berlin (and Germany in general) also lacks robust disaster resilience, due to fragmented governance at the federal and subnational levels.[11] This almost certainly hinders coordinated action on maximizing the synergies from collaborative, cross-sector mitigation, and adaptation.

Moreover, TMG's own data indicate that its renewable share in power consumption was 14.1% in 2017. This is admittedly a significant increase over 8.7% in 2014 when TMG was incentivized to harvest a lot of low-hanging fruit. However, its

---

[8] A list of the over 3500 governments and other actors, plus selected "success stories," is available at "We Are Still In": https://www.wearestillin.com.

[9] The U20 Group of Cities communique is available at the following URL: https://www.c40.org/press_releases/urban-20-group-of-cities-meet-in-tokyo-and-urge-g20-to-act-urgently-on-climate-change-social-inclusion-and-sustainable-economic-growth.

[10] The data are outlined in Jess Smee, "Energy use in the city of Berlin," Clean Energy Wire, January 16, 2019: https://www.cleanenergywire.org/factsheets/energy-use-city-berlin.

[11] See the comments by German government officials in Ben Knight, "Berlin blackout raises questions over Germany's power grid," DW.com, February 28, 2019: https://www.dw.com/en/berlin-blackout-raises-questions-over-germanys-power-grid/a-47730394.

official 2016–20, JPY 270 billion "smart energy city" plan projects the renewable share to increase to 15% by 2020%, 20% by 2024%, and 30% by 2030, dependent in large part on increased efficiency (from LEDs and other measures) that reduces energy demand by 30% by 2030% (vs. 2000 levels).[12]

In short, TMG's official energy plan shows that 100% renewable power by 2030 is very unlikely. Moreover, TMG's plans are considerably more sophisticated than, for example, New York City (a signatory to the U20 Group declaration), San Francisco, and other major US metropolises. Such cities routinely announce ambitious energy and emissions targets, to rounds of applause at conferences and in the press. However, unbiased surveys indicate that these cities rarely follow up with credible plans and programs. For example, the American Council for an Energy-Efficient Economy maintains a regularly updated "State and Local Policy Database," which reveals generally scant evidence of municipal renewable energy and energy-reduction goals in New York City and other major US urban centers.[13]

These are other problems with unduly expecting renewable energy to solve climate and other crises. Even the most ambitious projections by the International Renewable Energy Agency (IRENA) suggest that, at best, only 57% of power could be renewable by 2030. The report does not zero in on megacities. Moreover, it is marred by a lack of attention to the enormous governance changes needed to overcome NIMBY, restructure the incentives of vested interests, maximize cost–benefit performance from increasingly constrained critical materials (such as cobalt, copper, and lithium), and other issues.[14] So the IRENA's projections need to be viewed with a critical eye on how megacities (and other entities) can actually achieve the decarbonizing energy shifts that some deem technically possible. In addition, the Renewable Energy Policy Network for the 21st Century (REN21) "Renewable Energy in Cities: 2019 Global Status Report" emphasizes the climate, public health, socioeconomic, and other needs for cities to be aggressive on renewable energy and efficiency. It also points out that in 2018, fully one-fifth of global carbon-dioxide emissions originated from 100 cities, such as London, New Delhi, Seoul, and other megacities. However, at the same time, the REN21 lament massive data gaps on municipal policies, generation capacities, investment, use of digital technologies,

---

[12] On these data, see (in Japanese) the "Smart Energy City" section of TMG's energy plan: https://www.seisakukikaku.metro.tokyo.jp/basic-plan/actionplan-for-2020/plan/pdf/honbun2_smartcity_1.pdf.

[13] The American Council for an Energy-Efficient Economy "State and Local Policy Database" is available at the following URL: https://database.aceee.org.

[14] See "Global Energy Transformation: A Roadmap to 2050," International Renewable Energy Agency, 2019: https://www.irena.org/publications/2019/Apr/Global-energy-transformation-A-roadmap-to-2050-2019Edition.

and other core elements of credible targets.[15] This evidence further calls into question the U20 Group of Cities' commitment to 100% renewable power by 2030.

For its part, the International Energy Agency's (IEA) ambitious "Sustainable Development Scenario" in the 2018 World Energy Outlook (WEO 2018) forecasts at best about half of the 2030 global generation total of over 30,000 TWh (vs. roughly 26,700 TWh in 2018) derived from renewables.[16]

Good governance to manage multiple challenges

Moreover, the IEA does pay increasingly close attention to material constraints. Recent releases highlight that ambitious policies on renewables and electric mobility imply cobalt, lithium, nickel, and other critical material demand that exceeds current supply.[17] The IEA's concerns parallel those of the Japanese,[18] the European Union,[19] the California Business Roundtable,[20] and a steadily growing number of other actors. Many of these materials are used at far greater density, per unit of energy consumption or production, in green technologies as compared to conventional power systems, automobiles, and the like. Moreover, supplies of these materials have other competing sources of demand, including smart phones, jet engines, healthcare, and multiple other areas. The IEA and other analyses discuss supply constraints, geostrategic risks, human rights concerns, environmental damage (from harvesting and processing critical materials), and other issues. These challenges are all central to sustainable development and the circular economy. The emerging facts suggest that any credible, rapid shift to sustainable energy and efficiency will require prioritizing the use of constrained critical materials. Doing that will almost certainly require TMG-style collaborative and integrated governance.

A further item overlooked by the U20 Group's ambitious commitment is recent evidence of the enormity of the "Global Cooling Challenge" in the context of climate change, population increases, and expanding urbanization. Although air

---

[15] See the preliminary findings in the "Renewable Energy in Cities: 2019 Global Status Report," Renewable Energy Policy Network for the 21st Century (REN21): http://www.ren21.net/cities/wp-content/uploads/2019/05/REC-GSR-2019-Preliminary-Findings-final.pdf.

[16] The "World Energy Outlook 2018" scenarios developed by the International Energy Agency are available at the following URL: https://www.iea.org/weo/.

[17] See, for example, "Global EV Outlook 2019," International Energy Agency, May 27, 2019: https://www.iea.org/publications/reports/globalevoutlook2019/.

[18] Japan's JOGMEC and other agencies produce a range of materials, as do the carmakers (e.g., Toyota), battery suppliers (e.g., Panasonic), metal firms (e.g., Mitsubishi Materials) and other concerns.

[19] See, for example, EURACTIV's November 2018 work on "Metals in the circular economy": https://www.euractiv.com/section/circular-economy/special_report/metals-in-the-circular-economy/.

[20] See "A Closer Look at California's Cobalt Economy," California Business Roundtable, January 2019: https://centerforjobs.org/wp-content/uploads/A-Closer-Look-At-Californias-Cobalt-Economy-2.pdf.

conditioning is taken for granted in much of the developed world, it is critical to development. Singapore's founder, Lee Kuan Yew, is famous for having ranked air conditioning alongside interethnic tolerance as foundations of development. Yew asserted that:

Air conditioning was a most important invention for us, perhaps one of the signal inventions of history. It changed the nature of civilization by making development possible in the tropics.

Without air conditioning you can work only in the cool early-morning hours or at dusk. The first thing I did upon becoming prime minister was to install air conditioners in buildings where the civil service worked. This was key to public efficiency.[21]

Moreover, air conditioning is especially crucial to human health in the midst of rising heat and humidity and increasingly frequent heat waves. The current global average in use of air conditioning is 720 hours/year. Due to climate differences, room air conditioner (RAC) usage hours per year in China average 545, in Japan 720, but 1600 in the United States. Usage equals or exceeds 1600 hours h/year in India, Mexico, Brazil, Indonesia, and the Middle East (the latter is an astounding 4672). Because of global climate change, these usage hours are increasing at an estimated 0.7%/year (leading to a 25% increase by 2050).

The Rocky Mountain Institute (RMI) and other partners, including many elements of the Indian Government (e.g., the Ministry of Power), have organized an initiative to cope with the unsustainable power demand posed by conventional air conditioning in a warming climate. The RMI analysts and their collaborators point out that the global number of RACs in 2016 was roughly 1.2 billion (over 400 million in China alone), and that this figure is likely to increase to 4.5 billion by 2050.

The RMI draw on IEA and other data indicating that supplying the power demand for this growth in RAC stock, much of which will be concentrated in growing global megacities, will require roughly USD 1.2 trillion in new generation capacity. This is because the 2016 global RAC power demand of 2300 TWh will likely be more than triple over the same period, reaching 7700 TWh in 2050 (about 16% of global electricity demand). That 5400 TWh increase in power demand between 2016 and 2050 would require an astounding addition of 2000 GW of generation capacity, equivalent to "the current annual electricity consumption of the US, Japan, and Germany combined." Moreover, the cumulative GHG emissions (from power demand as well as the effect of refrigerant gases), projected at between 132 and 167 GT, would likely exhaust 25%–50% of the remaining carbon budget.

---

[21] Cited in Katy Lee, "Singapore's founding father thought air conditioning was the secret to his country's success," Vox, March 23, 2015: https://www.vox.com/2015/3/23/8278085/singapore-lee-kuan-yew-air-conditioning.

In India alone, where RAC penetration is only 7% but sales are already increasing at 15%/year, the RMI and IEA estimates indicate a more than 20-fold increase in power demand for RAC, from 94 TWh in 2016, to 1890 TWh in 2050. Seen in per-capita terms, urbanization in India is projected to raise RAC demand from a current global low of 72 kWh to 1140 kWh. Satisfying that level of demand would require India to install fully one-third of the global 2000 GW of needed new generation capacity. The RMI is a staunch advocate of renewable energy and efficiency. Hence, it is not deliberately bearish in warning that "[w]e cannot solve this magnitude of growth by adding renewables alone." It points out that in 2017 the total global increase of 94 GW in solar generation capacity was less than that year's RAC incremental demand growth of 100 GW.[22]

All these data were available in advance of the U20 Group of Cities' declaration, so it is mystifying that they were ignored. Given the gravity of climate change and other threats to sustainability, the political optics of ambitious declarations should not take precedence over concrete and credible policymaking. Megacity leadership on sustainability will require hard choices that maximize the synergies from mitigation and adaptation to leave sufficient material, fiscal, and other resources for equity, health, and other sustainability goals. Recent scholarship is advancing the argument that robust policy tools may have to look to wartime rationing as a model.[23] So it is especially ironic and regrettable that global megacity leaders met in Tokyo and released a communique that completely ignored TMG's essential governance lessons.

TMG in comparative perspective

TMG—and Japan as a whole—is marked by considerably less stress on bold energy and emissions targets. Even the Japanese version of the U20 Group of Cities' declaration omits the 100% renewable power by 2030 commitment, emphasizing instead carbon neutrality and 100% renewable energy by 2050. It also stresses the role of robust governance and collaboration in achieving sustainable development goals (SDGs), resilient infrastructure, and inclusivity.[24] Rather than emphasizing attention-getting targets, TMG and Japan evince a great deal of more concern to

---

[22] See Ian Campbell et al., "Solving the Global Cooling Challenge," The Rocky Mountain Institute, November 2018: https://rmi.org/wp-content/uploads/2018/11/Global_Cooling_Challenge_Report_2018.pdf.

[23] On this see Peck D.P. et al., "Product policy and material scarcity challenges: the essential role of government in the past and lessons for today," in Bakker CA and Mugge R eds. PLATE: Product Lifetimes and The Environment, IOS Press, 2017: https://www.iospress.nl/book/plate-product-lifetimes-and-the-environment/.

[24] See (in Japanese) "Main points of the 2019 U20 Tokyo Mayors' Summit Communique," Tokyo Metropolitan Government, May 22: http://www.metro.tokyo.jp/tosei/hodohappyo/press/2019/05/22/documents/04_01.pdf.

expand the scope and scale of planning to confront a multiplicity of unprecedented challenges. Japan's approach indicates that building up resilient adaptation to all hazards is a cornerstone of multistakeholder agreement on mitigation and other public goods. In that context, TMG is an important nexus of multilevel governance that offers potent lessons for how megacities can actually implement technological solutions for sustainability.

Of course, no national or subnational entity on the planet is doing enough, in mitigation and adaptation, to address the severity of climate and other threats. That said, TMG is a sustainability leader when seen in comparative light. As a megacity, Tokyo's per-capita energy, water, and waste flows are considerable below the average of peers such as Shanghai, New York City, London, Paris, and others.[25] TMG also installed Japan—and the world's—first urban cap and trade scheme that includes the commercial and industrial sector, "including office buildings, which are often concentrated in megacities."[26] Indeed, TMG and other Japanese cities are generally built in a compact manner that fosters efficient resource use. One recent recognition of this is seen in the UNEP's International Resource Panel (IRP) 2018 report on "The Weight of Cities: Resource Requirements of Future Urbanization." The report notes that "Japanese cities have the densest and most connected street patterns," with Tokyo's "level of transit connectivity and intensity of use" being the highest in the world. Because of these structural factors, "Japan has the highest world energy productivity (ratio of energy consumption to added value), close to three times the global average."[27] The Tokyo Metro transit network (primarily subway) reflects this: its 382 km length is far less than Shanghai's 639 km, and even New York's 401 km; however, in 2018, its annual ridership of 3.463 billion was the world's largest, dwarfing that of second-ranked Moscow (2.369 billion), third-ranked Shanghai (2.044 billion), and sixth-ranked New York City (1.806 billion).[28]

TMG's commitment to increasing this level of sustainability is seen in its FY 2019 budget, which features a three-tiered approach to maximizing relevant amenities. These three are "safety city," "smart city," and "diver-city" (the latter a combination, in Japanese script, of "diverse" and "city"):

---

[25] These results were reported by what appears to be the first ever comparison of energy and other resource flows in megacities. See Kennedy, C. et al. "Energy and material flows of megacities," Proc Natl Acad Sci U S A. 2015 May 12: https://www.ncbi.nlm.nih.gov/pmc/articles/PMC4434724/.

[26] See "Creating a Sustainable City: Tokyo's Environmental Policy," Tokyo Metropolitan Government, September 2018: http://www.kankyo.metro.tokyo.jp/en/about_us/videos_documents/documents_1.files/creating_a_sustainable_city_2018_e.pdf.

[27] See p. 109 "The Weight of Cities: Resource Requirements of Future Urbanization," International Resource Panel, 2018, available at the following URL: http://www.resourcepanel.org/reports/weight-cities.

[28] The data are compiled by the International Association of Public Transport, and published as "World Metro Figures 2018": https://www.uitp.org/sites/default/files/cck-focus-papers-files/Statistics%20Brief%20-%20World%20metro%20figures%202018V4_WEB.pdf.

(1) The safety-city budget centers on a large investment of JPY 300 billion in disaster-resilient water networks, carbon-sequestering levees, and other items.
(2) The smart city component of the budget totals JPY 326 billion. The largest portion (JPY 207 billion) of this spend focuses on building up transport efficiency by integrating telework, transit-demand management, and other smart initiatives.
(3) The diversity-city aspect totals JPY 353 billion. Of this, fully JPY 174 billion is devoted to augmenting the national government's initiatives in providing free child-care and kindergarten education. At first glance, these items may seem extraneous to climate objectives. However, relieving child poverty, increasing women's opportunities, and enhancing work-life balance are in fact crucial elements of sustainability.[29]

In addition, TMG has been using the impending 2020 Olympics as a deadline and venue for realizing multiple sustainability projects. One is the Tokyo 2020 Medal Project, which aims to produce the Olympic medals from recycled materials. For this objective, by January of 2019, 67,180 tons of discarded mobile phones and other electronic devices were collected by 1594 participating local governments (via, e.g., 18,000 discard boxes). Moreover, collections via agreement with the major mobile carrier NTT Docomo resulted in the recovery of 5.75 million used mobile phones. These will supply 93.7% of the gold (30.3 kg), 85.4% of the silver (4100 kg), and 100% of the bronze (2700 kg) needed.[30] Moreover, the project also normalizes recycling to recover critical materials.

The drivers for megacity sustainability

We shall examine in more detail later how TMG invests in smart city initiatives, building on its compact, resource-efficient advantage. However, first, we survey the drivers for its action.

One prominent driver of the TMG sustainable megacity is institutionalized recognition of the need to cope with massive numbers. We have seen that TMG's population, economic output, and other metrics are enormous. These facts are the fruits of planning, and making them more sustainable will also largely derive from smarter planning.

TMG's sustainability planning embraces the scale of the global climate and resource crises. These numbers include population increases, higher levels of urbanization, and massive material requirements. For example, the global population has risen from 3 billion in 1960 to 7.7 billion in 2019 has passed the 7 billion mark and

---

[29] Concerning TMG's FY 2019 budget, see (in Japanese) the succinct summary at Tokyo Metropolitan Government News, March 1, 2019: http://www.koho.metro.tokyo.jp/2019/03/documents/201903.pdf.
[30] The project and its status are detailed (in Japanese) at "the progress of the project," Tokyo 2020: https://tokyo2020.org/jp/games/medals/project/status/.

appears likely to reach 10 billion by 2050.[31] Over the same period, the rate of urbanization has mushroomed. The United Nations Department of Economic and Social Affairs, Population Division, compiles surveys of "World Population Prospects." The Division's May 16, 2018 revision and subsequent updates indicate that global urbanization was roughly 30% in 1950, but had risen to 55.3% in 2018. It projects that this ration of urbanization is likely to increase to 60% by 2030 and then roughly 66% by 2050. It projects that most of this increase will be concentrated in the Asia–Pacific, whose megacities (10 million or more residents) are expected to swell from 20 in 2018 to 27 in 2030.[32]

The increase in total global population, together with the share living in cities, has profound implications for sustainability. We saw that earlier regarding global cooling needs. The United Nations Environmental Programme (UNEP) has warned that cities already "consume 75% of the world's natural resources, 80% of the global energy supply and produce approximately 75% of the global carbon emissions."[33] Imagine if ongoing development in India and elsewhere proceeds along conventional lines. The OECD's 2015 report on Material Resources, Productivity, and the Environment revealed that, in 2011, the average, daily per-capita consumption of materials in OECD countries was as follows: 10 kg of biomass, 18 kg of construction and industrial minerals, 13 kg of fossil energy carriers, and 5 kg of metals.[34] The UNEP's International Resource Panel (IRP) is equally stark in its 2018 report on "The Weight of Cities: Resource Requirements of Future Urbanization." The IRP working group on cities undertook the assessments, and determined that continued conventional urbanization would increase annual urban resource requirements from 40 billion tonnes in 2010 to 90 billion tonnes in 2050.[35]

Adaptation and mitigation synergies

An additional factor underpinning planning is that TMG is very vulnerable to disasters. As is generally well known, TMG confronts serious seismic risks. Japan as a whole represents only 0.3% of the terrestrial surface, but is the site of roughly 20% of the world's large earthquakes and 10% of the most active volcanoes. Greater

---

[31] See the May 2019 revised assessment of "World Population Growth" by Max Roser, Hannah Ritchie and Esteban Ortiz-Ospina in Our World in Data: https://ourworldindata.org/world-population-growth.

[32] The report and summaries are available at the following URL: https://www.un.org/en/development/desa/population/theme/urbanization/index.asp.

[33] See United Nations Environmental Programme, "Cities and Buildings," (nd):http://www.unep.org/SBCI/pdfs/Cities_and_Buildings-UNEP_DTIE_Initiatives_and_projects_hd.pdf.

[34] See OECD, Material Resources, Productivity and the Environment, Paris: OECD, February 2015, p. 9.

[35] The International Resource Panel's report is available at the following URL: http://www.resourcepanel.org/reports/weight-cities.

Tokyo rests on three separate plates, directly above the North American Plate, beneath which is the Philippine Sea Plate, under which is the Pacific Plate. The probability of a magnitude 6 or greater quake over the coming 30 years varies from 48% in TMG per se to 85% in Chiba City, the easternmost extent of greater Tokyo. The entire region was devastated by the 1923 Great Kanto Earthquake, whose magnitude 7.9 shock killed roughly 100,000 people and gave rise to a conflagration that devastated 3470 ha (44% of then Tokyo City). The region's main reconstruction initiatives continued until 1930, and contributed greatly to the development of urban planning in Japan.[36]

In addition to seismic risks, Japan is also disproportionately threatened by typhoons, intense rain, heat stress, and other phenomena worsened by accelerating climate change. That multiplicity of hazards means that TMG and other Japanese cities, like much of Asia, confront the risk of multiple hazards erupting simultaneously, to produce cascading failures across a broad range of critical infrastructures.

Fig. 13.2 displays Japan's flood risks, showing that half the population and three quarters of the assets are concentrated into the 10% of land that is coastal flood plains.[37]

The topography of Tokyo is particularly fraught. Fig. 13.3 displays the Sumida, Ara (in the figure, "Arakawa"), Edo, and other rivers that run through several of TMG's cities and wards, such as Arakawa-ku, Adachi-ku, and Misato City. The rivers—especially the Edo River—are already considerably higher than their surrounding districts. As sea levels continue to rise, these river levels also rise. Moreover, as bouts of intense rain become more common, the threat of catastrophic flooding correspondingly increases.

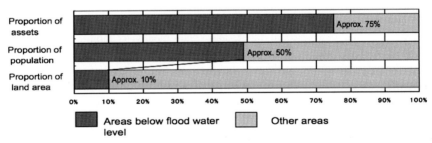

FIGURE 13.2

Japan's flood risk.

Adapted from Atsumi, 2009.

---

[36] On these matters, see "The Study of Reconstruction Processes from Large-Scale Disasters," Japan International Cooperation Agency, November 2013: https://www.jica.go.jp/english/news/focus_on/c8h0vm00008lxw0n-att/process_01.pdf.

[37] See Atsumi, M. "River management in Japan," River Bureau, Ministry of Land, Infrastructure, Transport and Tourism, Japan, January 2009: https://www.mlit.go.jp/river/basic_info/english/pdf/conf_05.pdf.

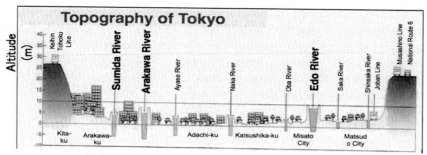

FIGURE 13.3

Tokyo's flood risk.

Adapted from Atsumi, 2009.[38]

So geography and hydrology are facts that shape TMG's planning, and encourages its regionalization. Fig. 13.4 affords a generalized perspective on the items elucidated in Figs. 13.2 and 13.3. It offers a visual representation of Japan, which is roughly 70% mountainous, with half the population and three quarters of the assets concentrated into the 10% of land that is coastal flood plains. Japan thus has short and steep rivers, compared to the Seine, Rhine, Mekong, and other major global rivers. Japanese rivers' short length and steep pitch derive from the fact that the country is quite mountainous and narrow (no point in Japan is more than 150 km from the sea). Moreover, Japan receives roughly twice the global annual average of rainfall. These factors make the flow variability of Japan's rivers unusually great. As has been said for over a century, Japanese rivers are more like waterfalls during the country's seasonal bouts of concentrated rainfall.

The figure is from 2007, when TMG and the national government were using cameras, sensors, conventional radar, and other means to monitor and control dam, river, and urban water assets to limit flood and other risks. A more recent figure (not currently available) would show that these monitoring and control plans have advanced considerably. This is in large part because of the subsequent evolution of information technology (ICT, IoT, AI); the cheapening of sensors for monitoring water height, flow, and other parameters; and the ongoing integration of dam-asset planning with planning for other critical infrastructures.

Smart technology is used in the development and deployment of advanced radars for bolstering meteorological data against extreme weather. These next-generation radars give rapid and pinpoint advance warning of impending rainfall. That situational awareness allows water managers to adjust dams, river protections, sewerage systems, and other critical infrastructures to cope with the hydrological challenges.

---

[38] See Atsumi, M. "River management in Japan," River Bureau, Ministry of Land, Infrastructure, Transport and Tourism, Japan, January 2009: https://www.mlit.go.jp/river/basic_info/english/pdf/conf_05.pdf.

FIGURE 13.4

Japan's topography.

Source: MLIT 2007.[39]

TMG's sewerage division has already deployed this integration of technology. TMG uses advanced radar and monitoring technologies to manage its 16 million meters of pipes that move 2.2 million cubic meters of water per day. At the same time, TMG maximizes the deployment of renewable energy in the system, the efficiency of pumps and other equipment, and the recovery of such valuable resources as phosphorous (crucial to agriculture). The radar networks also afford valuable information in other areas of the economy, such as energy management, agriculture, retail, tourism, and the myriad other domains where reliable weather data are useful. Moreover, TMG quite proudly publicizes its use of advanced radars and other technologies, regularly producing quite readable pamphlets as well as very professional videos for the public. The public education helps to reinforce the link between resource-efficient adaptation and mitigation.

TMG residents thus express strong support for disaster resilience, is seen in Table 13.1. The table reproduces the most recent results of TMG's annual survey

[39] See (in Japanese) MLIT (2007) "Looking at Dam Projects," Ministry of Land, Infrastructure, Transport and Tourism (MLIT), Japan: http://www.mlit.go.jp/river/pamphlet_jirei/dam/gaiyou/panf/dam2007/pdf/4-2.pdf.

Table 13.1 Tokyo metro resident survey of priorities, 2010–18.

Rank	2010	2011	2012	2013	2014	2015	2016	2017	2018
1st	Pub. Saf. (50.7)	DRR (53.4)	DRR (54.9)	DRR (52.7)	DRR (49.4)	Aging (49.8)	Aging (53.5)	DRR (48.7)	Aging (53.2)
2nd	Aging (49.4)	Aging (43.7)	Pub. Saf. (47.5)	Pub. Saf. (48.1)	Pub. Saf. (47.7)	Pub. Saf. (47.7)	DRR (48.6)	Pub. Saf. (48.2)	DRR (51.3)
3rd	Med. (49.0)	Pub. Saf. (43.6)	Aging (43.6)	Aging (44.2)	Aging (46.5)	Med. (41.9)	Pub. Saf. (48.1)	Aging (46.7)	Pub. Saf. (46.5)
4th	DRR (33.0)	Med. (40.8)	Med. (41.5)	Med. (38.0)	Med. (43.1)	DRR (41.6)	Med. (41.7)	Med. (41.5)	Med. (41.0)
5th	Enviro (32.0)	Enviro (28.2)	Enviro (25.8)	Enviro (27.3)	CA (26.5)	Trans. (23.1)	Admin (27.1)	Admin. (31.2)	Admin (27.2)

Admin. = Administration and Finance; CA = Consumer Affairs; DRR = Disaster Risk Reduction; Enviro. = Environment; Med. = Medical Services and Sanitation; Pub. Saf. = Public Safety; Trans. = Transportation.
Source: TMG, 2018.

of 3000 residents concerning the priorities they wish TMG to emphasize. The poll offers 30 items from which respondents are asked to select 5 in terms of priority. The results indicate that disaster risk reduction (DRR) was ranked #1 from 2012 to 2014, then dipped to #4 in 2015, but rose again to second place in 2016. In 2017, it returned to the top, just above public safety (i.e., policing, etc.). In 2018, DRR scored slightly below aging, but still ranked second. Environmental measures (Enviro) routinely score far lower, being sixth in 2018 (chosen by 24.4% of respondents as a priority).

This survey evidence indicates that TMG residents strongly support spending on resilience, much more than explicitly environmental measures. This is one reason to look for how policymakers link their energy projects to the "resilience" theme. In TMG, the approach is evident in the use of the sewerage network as an antiflood system (integrated with advanced radar and supercomputing) as well as a means for boosting TMG energy and cutting its GHG emissions (sewer operations result in 35% of TMG greenhouse gas emissions).

TMG's FY 2018 budget projected spending JPY 88.7 billion on the smart city, an increase of JPY 16 billion over the previous year. Given the earlier, it is no surprise that mitigation measures (including EVs, solar, hydrogen stations) were budgeted at JPY 12.7 billion while adaptation measures (bolstering river levees) were budgeted at JPY 76 billion. We saw earlier that these measures are significantly increased in FY 2019.

In short, the evidence suggests that TMG's residents are quite amenable to changing the built environment as an adaptation response. That in turn, allows for significant mitigation via the rollout of renewable energy, smart networks, and compact communities to enhance resilience against disasters, demography, and a multiplicity of other challenges.

The global challenge

As noted earlier, TMG is a central player in a national regime of building resilience in the face of all hazards, including the multiple manifestations of climate change. Let us turn to consider how that approach maps with global needs, by examining disasters and their costs.

Comprehensive global summaries of disasters and their costs are available from the United Nations Office for Disaster Risk Reduction (UNDRR). As of this writing, their most recent work is a report on "Economic Losses, Poverty and Disasters, 1998—2017."[40] The report summarizes total reported disaster losses, assessing their direct economic affect during the 1998—2017 period as in excess of USD 2.9 trillion, with climate-related disasters representing 77% of the total, or roughly USD 2.25 trillion. The report compares these aggregate figures with the 1978—97 period,

---

[40] The report can be accessed at the following URL: https://reliefweb.int/report/world/economic-losses-poverty-disasters-1998-2017.

when total reported losses were far less, at just over USD 1.3 trillion with climate-related disasters accounting for 68%, or USD 895 billion. It also points out that there is serious underreporting of economic costs. Over the 1998−2017 period, high-income countries reported economic costs for 53% of disasters, while low-income countries reported assessments for only 13% of disasters.

Fig. 13.5 displays the top 10 countries/regions in absolute, direct economic losses, dividing these countries/regions into high income, upper-middle income, lower-middle income, and low income. As is indicated, the greatest absolute economic losses afflict the United States, with storm damage ranking highest in USD 944.8 billion in damages over the 1998−2017 period. This amount is followed by China, where flood damages front a cumulative total of USD 492.2 billion.

It is also important to note that the UNDRR report aggregates reported disaster losses of all types. These disasters include seismic events that cannot be attributable to climate change, at least in the present.[41] These latter were responsible for only 9% of disasters. However, including them is key to formulating an all-hazard approach to enhancing public safety. The report shows that during the 1998−2017 period, reported deaths total 1.3 million, with 4.4 billion people having been injured, made homeless, displaced or requiring emergency assistance. Seismic events led to the greatest loss of lives, as a total of 563 earthquakes, including the tsunamis they generated, caused 56% of total deaths, representing 747,234 lives lost.

Fig. 13.6 from the UNDRR report shows that over the 1998−2017 period the most frequent disaster was floods, at 43.4% of reported events, followed by storms, at 28.2%. The greatest absolute economic losses afflict the United States, with storm damage ranking highest in USD 944.8 billion in damages over the 1998−2017 period. This amount is followed by China, where flood damages front a cumulative total of USD 492.2 billion.

The cost of disasters as a share of GDP looks different. The high-income countries largely disappear from the list, with the exception of the US territory of Puerto Rico, whose disaster costs are a prodigious 12.2% of GDP. The highest, however, is Haiti, where assessed and reported disasters impose a crushing burden of 17.5% of GDP. Hence, with the exception of Puerto Rico, the countries/regions with the highest relative disaster burdens are lower-middle and low-income. Moreover, although most of the disasters are floods and storms, seismic events also play a large role.

Fig. 13.7 shows that over the period surveyed, storm damages made up a rough average of 60% of climate-related disaster losses. The definition of storms includes tropical cyclones and hurricanes. Not only was the economic damage significant, but the death toll was as well. The report finds that storms were the second-leading cause of death over the period, costing 233,000 lives.

---

[41] Note, however, that a significant scientific literature explores the relationship between seismic events and ice-mass loss, methane-hydrate melt and other phenomena. The researchers study such changes in the past in part to investigate whether current climate processes may induce seismic activity (Brandes, 2018; McGuire, 2012).

Top 10 countries/territories for cumulative losses compared to
top 10 countries/territories for losses relative to GDP 1998–2017

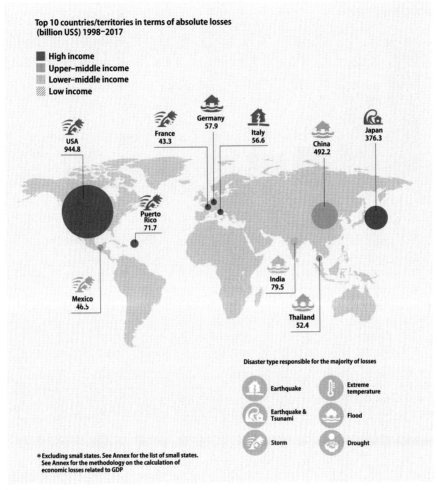

Top 10 countries/territories in terms of absolute losses
(billion US$) 1998–2017

■ High income
■ Upper–middle income
▨ Lower–middle income
▨ Low income

Germany
57.9

France
43.3

USA
944.8

Italy
56.6

China
492.2

Japan
376.3

Puerto
Rico
71.7

Mexico
46.5

India
79.5

Thailand
52.4

Disaster type responsible for the majority of losses

Earthquake

Earthquake &
Tsunami

Storm

Extreme
temperature

Flood

Drought

* Excluding small states. See Annex for the list of small states.
See Annex for the methodology on the calculation of
economic losses related to GDP

FIGURE 13.5

Disasters and absolute cumulative losses, 1998–2017.

Source: Wallemacq and House, 2018.

However, the aggregate losses from disasters are underreported. Fig. 13.8 is one indicator of how much the aggregate losses from disasters are underreported. It shows that economic costs for storms were reported in just over half of all cases. It also reveals that the direct economic costs of other climate-related phenomena (such as wildfires, floods, and droughts) were reported in far less than half of all cases. Indeed, the costs of extreme temperatures were assessed and reported in only 11% of such incidents. One can infer that total economic costs of disasters would be considerably higher if there were more accurate assessment (particularly

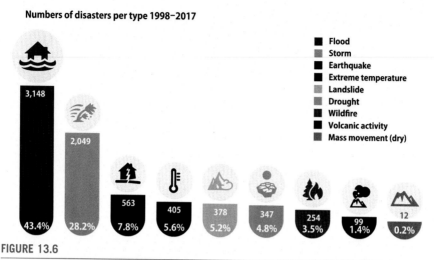

FIGURE 13.6

Disasters by type, 1998–2017.

Source: Wallemacq and House, 2018.

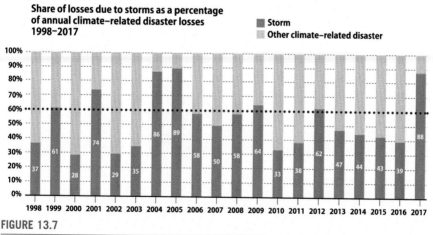

FIGURE 13.7

Storms as a share of climate-related disasters, 1998–2017.

Source: Wallemacq and House, 2018.

the inclusion of indirect economic costs, such as long-term impacts on human health).

Table 13.2 provides additional evidence of this underreporting, in this case breaking it down by country/region income level. The discrepancy between high-income and low-income reporting of losses is striking, with the former being 53% of all disasters whereas the latter are only 13%.

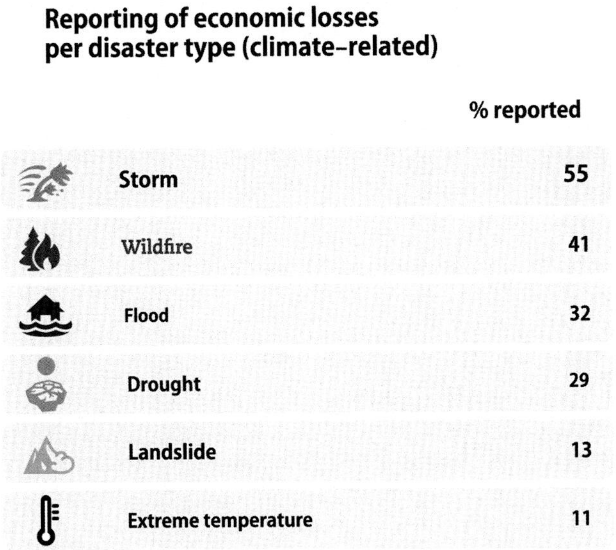

Reporting of economic losses
per disaster type (climate–related)

		% reported
🐉	**Storm**	**55**
🔥	Wildfire	41
🏠	Flood	32
	Drought	29
	Landslide	13
🌡	Extreme temperature	11

FIGURE 13.8

Reporting of economic losses, 1998–2017.

Source: Wallemacq and House, 2018.

Table 13.2 Reporting of economic losses, by income level, 1998–2017.

Income	all disasters	climate-related	geophysical
High	53	52	61
Upper-middle	40	40	37
Lower-middle	31	30	31
Low	13	13	20

Source: Wallemacq and House, 2018.

Fig. 13.9 displays continental variations in the frequency of occurrence of geophysical disasters, deaths, displacement and other effects, and economic losses. The Asian region is shown to be subject to most impacts. These geophysical events cannot be attributed to climate change, at least in the present. However, one could argue that their impacts weaken the resilience of individuals and communities, rendering them more vulnerable to climate-driven disasters as well as slow-onset effects (such as increasing humidity or aridity, sea-level rise, and so on). In this respect, it is important to keep in mind that the Asian region is home to 4.5 billion people, or about 60% of the global population of 7.6 billion.

Disasters drive internal displacement, and displacement undermined sustainable development. The Internal Displacement Monitoring Centre (IDMC) highlights this reciprocal interaction. The IDMC was established in 1998, to provide accurate data

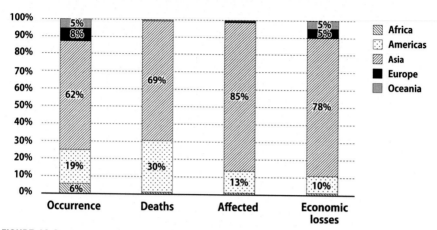

Relative human and economic costs of geophysical disasters on continents 1998–2017

FIGURE 13.9

Disaster impacts, by continent, 1998–2017.

as well as analyses based on it. In addition to research papers and timely information, it publishes an annual global report that synthesizes estimates of displacement due to conflict and disasters.[42] As of this writing, the IDMC's most recent global report is the "Global Report on Internal Displacement 2018" (hereafter, "GRID 2018"), released in May 2018. The report provides excellent graphical summaries of the scale of displacement and conflict. Its data indicate that in 2017 an additional 30.6 million people were internally displaced by disasters and conflict. Of that total, the number displaced by disasters was 18.78 million, whereas 11.77 million were displaced by conflict. The data also show that disaster displacement is largely clustered in the Asia—Pacific while conflict displacement is centered in Africa and the Middle East. We see the consequences in Fig. 13.9.

Because of complexity, an integrated approach to resolving or at least alleviating the crises is needed. As highlighted in Fig. 13.10, SDGs are focused on this integration. The figure shows that all of the goals are affected by internal displacement and the goals in turn affect internal displacement. Have to accelerate moves toward bolstering SDGs through maximizing cobenefits, which is already happening in governance and business (due to impacts on communities, supply chains).

New research on resilient infrastructure backs up this view. It highlights resilient infrastructure as crucial to realizing SDGs and helping people stay in place or move from rural areas to safer, healthier cities. These are no-regret pragmatism: in the face of academic uncertainty about the precise links between climate change, migration,

---

[42] The Internal Displacement Monitoring Center describes its role at the following URL: http://www.internal-displacement.org/about-us/.

Internal displacement and the SDGs

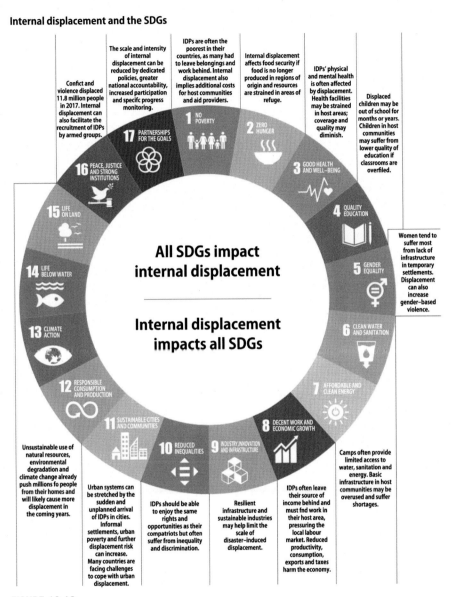

FIGURE 13.10

Internal displacement and sustainable development goals. IDMC: 2018.

and conflict, there can be no doubt that providing children with clean water and other public goods will alleviate a lot of health and other hazards.

For example, in October 2018 Oxford University and the United Nations Office for Project Services produced a report on the crucial role of resilient infrastructure, titled "Infrastructure Underpinning Sustainable Development."[43] They place infra-structure investments in a broader mix of critical public goods. These latter include investments in health, education, and food security, realized by climate-smart agriculture and sustainable management of water resources. They also highlight the role of renewable energy and other infrastructure.

TMG's example suggests that robust and integrated governance can help cities maximize mitigation and adaptation. Even megacity governments cannot do this on their own, so it requires the "all of government" approach suggested by the UN Habitat.[44] It also requires international collaboration, as seen in the initiatives by the C40 Cities Climate Leadership Group, the ICLEI (Local Governments for Sustainability),[45] and other international actors. The crucial and expanding role of the public sector, on the international stage, is also evident in potential Japan–China collaboration on building disaster-resilience into the Belt and Road initiative.[46]

TMG also affords a notable example of realizing adaptation and mitigation simultaneously in the smart megacity. Although "smart city" is often derided as a buzzword, and often indeed lacks real content, defining and standardizing the paradigm became an international priority from the mid-2010s. Since February 2014, a promising initiative to help define and develop truly smart and resilient cities has been undertaken by the International Standards Organization Technical Management Board's (ISO TMB) Smart Cities Strategic Advisory Group. The ISO has 165 member countries, representing 97% of the global population. As an agency, the ISO develops international standards for products, coordinating the product standards of its member countries' national standards bodies. Reflecting the concern that "smart city" and its component technologies lacked clarity and credibility in spite of rapidly accelerating climate and other crises, the ISO has set up the Smart Cities Strategic Advisory Group in February 2014 and mandated it to report in September 2015.

The ISO Group sought to work out a definition and approach based on cities' needs and how to address them. Putting the climate threat to the fore, the ISO Group contextualized the definition of "smart city" within the increasingly pertinent challenges of resilience and sustainability. To this end, the ISO Group conducted a "demand-side survey" of cities in 27 countries and consulted widely with experts and such international

---

[43] The report can be accessed at the following URL: https://www.itrc.org.uk/infrastructure-underpinning-sustainable-development/.

[44] In its May 31, 2015 issue paper on "Cities and Climate Change and Disaster Risk Management," the UN Habitat III argues that the "compactness, connectedness, inclusiveness and integration" principles of the New Urban Agenda can "improve disaster risk management, contribute to climate change mitigation and adaptation, as well as unlock opportunities for sustainable development": http://unhabitat.org/wp-content/uploads/2015/04/Habitat-III-Issue-Paper-17_Cities-and-Climate-Change-and-Disaster-Risk-Management-2.0.pdf.

[45] See "Innovation, collaboration and their citizens make cities smart," ICLEI, November 4, 2014: http://www.iclei.org/details/article/innovation-collaboration-and-their-citizens-make-cities-smart.html.

[46] See (in Japanese) "Nikai speaks on the Belt and Road initiative," Nihon Keizai Shinbun, April 26, 2019: https://www.nikkei.com/article/DGXMZO44263190W9A420C1EA3000/.

bodies as C40, ICLEI, UNEP, Smart Cities Council, the OECD, and others.[47] As the ISO Group pointed out in its September 2015 final report, they "recognized the complex structure of systems of systems that smart cities requires (e.g., water systems, energy systems, mobility systems etc.)." They are well aware that cities are all unique, like the people who inhabit them, but they stress that all cities "share a common anatomy."

This common urban anatomy is the networks that move people, energy, waste, and water as well as provide light, security, and other essential services. Almost all of these networks are at present energy- and resource intensive. However, they are undergoing rapid technological innovation, as is evident in the rapidly falling cost of solar and ICT sensors, the diffusion of LED lighting, the ongoing implementation of 5G networks, and other developments, many of them mutually reinforcing.[48] TMG's example shows that making the urban anatomy compact and disaster-resilient gains public support and facilitates decarbonization.

Integrated resilience and the Japanese megacity paradigm

The Intergovernmental Panel on Climate Change's Fifth Assessment Report (hereafter, "IPCC AR5")[49] section on human security identifies "Critical infrastructure and state capacity" as a major concern. IPCC AR5 points out that "[c]limate change and extreme events are projected to damage a range of critical infrastructure, with water and sanitation, energy, and transportation infrastructure being particularly vulnerable. Climate change is expected to exacerbate water supply problems in some urban areas that in turn pose multiple risks to cities" (Adger, Neil W and Juan M. Pulhin et al., 2014)

Japan would appear to offer lessons in governance and the integration of technology, to help cope with the underlying crisis. For some observers, Japan seems an unlikely candidate to contribute to human security. Japan has long been engaged in bolstering human security through disaster risk reduction. The UNDDR's 2015—30 Sendai Framework on Disaster Risk Reduction[50] is heavily informed by Japanese expertise and experience. Moreover, Japan's JPY 5 trillion-plus program of National Resilience strongly expresses the governance and other goals of the Sendai Framework. Japan's

---

[47] The list of participants in the "demand-side survey" is available on page 6 of Francesco Dadaglio and Dave Welsh, "ISO Smart Cities - Key Performance Indicators and Monitoring Mechanisms," Paper Presented to ITU Forum on Smart Sustainable Cities, Abu Dhabi-UAE, 3—4 May 2015: http://www.itu.int/en/ITU-D/Regional-Presence/ArabStates/Documents/events/2015/SSC/S6-MrDWelsh_MrFDadaglio.pdf.

[48] For example, when ICT sensing technology is used to manage LED lighting systems, shutting them off when they are not needed. These systems can reduce lighting costs substantially.

[49] The Intergovernmental Panel on Climate Change's Fifth Assessment Report (AR5) was published in 2014 and is available here: http://www.ipcc.ch/report/ar5/.

[50] An overview of the United Nation's Office for Disaster Risk Reduction (UNISDR) Sendai Framework is available here: https://www.unisdr.org/we/coordinate/sendai-framework.

approach is hardly sufficient to cope with climate threats to human security. Japan affords some important, overlooked lessons in integrating hard and soft infrastructure for no-regrets solutions. Japan's approach maximizes the number of stakeholders and cobenefits, fostering pragmatic collaboration and bolstering human security.

Since 2014, Japan's imperative of resilient adaptation, for lifeline infrastructures (water, communications, transport), has become institutionalized in a variety of new commissions and agencies, including the National Resilience Promotion Office.[51] The policy is also inscribed in an expanding portfolio of national and subnational NRPs that have legal precedence over other plans.[52] This fact is outlined in Fig. 13.11, which shows the administrative hierarchy for planning. This is reproduced at the local level, such as in the NRP that TMG adopted in January 2016.[53]

By May 2019, all of Japan's 47 prefectural authorities, along with 192 cities and towns, had either adopted local versions of the NRP or were formulating plans.[55] Furthermore, these numbers are growing, fostered by local collaboration and other

FIGURE 13.11

National resilience planning in Japan.

Source: Japan Cabinet Secretariat (nd: 9).[54]

---

[51] An overview of some of the agencies and commissions is available (in Japanese) at: https://www.cas.go.jp/jp/seisaku/kokudo_kyoujinka/.

[52] The central government's National Resilience plans for 2014−18 are available (in Japanese) here: https://www.cas.go.jp/jp/seisaku/kokudo_kyoujinka/kihon.html.

[53] The TMG plan is available (in Japanese) at the following URL: http://www.metro.tokyo.jp/INET/KEIKAKU/2016/01/70q1k100.htm.

[54] Japan Cabinet Secretariat (nd) "Building National Resilience". http://www.cas.go.jp/jp/seisaku/kokudo_kyoujinka/en/e01_panf.pdf.

[55] Links to Japan's subnational National Resilience plans are available (in Japanese) here: https://www.cas.go.jp/jp/seisaku/kokudo_kyoujinka/tiiki.html.

means to diffuse the program and facilitate its adoption by cash-strapped and people-poor local governments.

The resilience budgets are also quite large. The initial budgets between FY 2014 and 2018 were over JPY 3.5 trillion.[56] In 2018, spending requests accelerated, driven by unprecedented disasters and other factors. The JPY 5.3 trillion request for FY 2019, when coupled with the inevitable supplementary budgets and expanding tax breaks,[57] will almost certainly see Japan spend much more on disaster-resilience than the JPY 5.3 trillion requested for national defense.[58]

Japan's National Resilience program is evolving into full-fledged industrial policy. The focus of National Resilience increasingly centers on information technology (ICT), the internet of things (IoT), and artificial intelligence (AI). It aims to smarten power, water, communications, transport, and other critical infrastructure as well as network them together. It is also deliberately linked to SDGs, to expand its potential as a means for international engagement. The most comprehensive and recent discussion of this use of smart technology is available (in Japanese) in Kashiwagi Takao's *Super-Smart Energy Society 5.0*, published on August 27, 2018. We saw earlier that TMG uses this policy and technology integration to cope with flood threats while also maximizing the cobenefits (renewable energy, recycling of phosphorous, and other materials).

As noted earlier, the IPCC AR5 report on human security identifies "critical infrastructure and state capacity" as a major concern. It is especially worried about hydrologic threats, through intense rain, drought, sea-level rise, and other hazards. Japan is increasingly good at linking the soft and hard infrastructures of resilience, through inclusive planning and networking critical infrastructures.

In fact, Japan's National Resilience institutionalizes the UNDRR's Sendai Framework. The Framework stresses the need for "prior investment," so as to build resilience in the face of multiple hazards and reduce their impact. It also argues for "mainstreaming disaster risk reduction," through an inclusive, whole of government approach that makes coping with hazards a priority in all planning initiatives. The Sendai Framework calls for "the full engagement of all State institutions of an executive and legislative nature at national and local levels and a clear articulation of responsibilities across public and private stakeholders, including business and academia, to ensure mutual outreach, partnership, complementarity in roles and

---

[56] The 2014–19 National Resilience budgets are available (in Japanese) here: https://www.cas.go.jp/jp/seisaku/kokudo_kyoujinka/yosan.html.

[57] A summary of the FY 2019 proposed tax reductions and exemptions is available (in Japanese) here: https://www.cas.go.jp/jp/seisaku/kokudo_kyoujinka/pdf/h31zeiseikaisei_gaiyou.pdf.

[58] Japanese defense spending, whose 1% of GDP level is low relative to the US (3.1%) and EU countries' general average (1.3%), gets overwhelming media attention, whereas its resilience investments are either overlooked or derided as pork barrel public works in international media and websites.

accountability and follow-up."[59] Japan's National Resilience is increasingly implementing that proactive integration of hard and soft infrastructures. It is also evolving a coordinated framework of central, regional, and local capacities and responsibilities in the face of what are unprecedented positive and negative externalities.

Japan's National Resilience is also iterative: it is annually updated and revised, in light of comparatively transparent and comprehensive performance targets. It is also publicly supported, responsive, collaborative, well-funded, and serves to unite innovative capacity on collective problems. National Resilience lacks explicit greenhouse gas mitigation targets, but uses the very real threat of natural disasters and other hazards to reshape energy, environmental, urban, fiscal, and related policy regimes. National Resilience has already led to broad collaboration among government agencies, the private sector and civil society. This collaboration is clear from the diverse involvement of NPOs, disaster professionals, local governments, business associations, and other stakeholders in drafting the national and local resilience plans. It is also evident in the composition of the 19 working groups that compile sectoral studies (on green infrastructure, fire prevention, landslide countermeasures, underground infrastructure mapping, and other items) within the Association for Resilience Japan.[60]

Climate change is a wicked problem and very likely an existential crisis. Assessing and addressing it is fraught with disagreements and distractions, and the example of air-conditioning shows that there are huge adaptation needs that are running into mitigation and material constraints. However, it seems coalitions can be built on the basis of including a variety of no-regret "solutions" framed in the SDGs. TMG's smart megacity initiatives and its incorporation into Japan's National Resilience paradigm suggest some avenues for coping with the country's own domestic challenges, while affording lessons and opportunities for more vulnerable countries and regions. There are few other concrete initiatives for bolstering the governance and integrating the critical infrastructures of global cities. So TMG offers valuable lessons for how collaborative governance and smart technology can maximize the effective use of constrained fiscal, material, human and other resources, as well as time.

---

[59] See United Nation's Office for Disaster Risk Reduction (2015), "Sendai Framework for Disaster Risk Reduction 2015–30," p. 13: https://www.preventionweb.net/files/43291_sendaiframeworkfordrren.pdf.
[60] The diverse membership of the Association's 19 working groups, together with reports and other details (such as meeting schedules), can be confirmed (in Japanese) here: http://www.resilience-jp.biz/wg/.

Conclusion

Woodrow W. Clark, II MA^3 PhD

Clark completed a book that focused on Circular Economics which will provide the background for economics becoming a science (see Q^2E, Nova Press, June 2019). Actual economics was never flat as Adam Smith and his linear economic theory of "supply and demand" with government being the "invisible hand." This entire theory has never been enacted or even proven. Instead there are enormous cases where economics is not linear ranging from new companies to technologies to finance. And governments are always involved in approaches to create jobs and businesses that pay taxes. This is what the theory of Western economics from Adam Smith did.

Above all, Circular Economics is both quantitative and qualitative (Q^2E) because finance, economics, business, government, investments and even nonprofits are both. Science is both Q^2E which economics needs to be. Q^2E are the same for legal and policy areas too, as well as public policy, international economics, and more. Finally, for economics to be a science using both Q^2E, it must "define" everything.

That is what linguistics did under the research and leadership of Professor Noam Chomsky at MIT in the late 1950s. Today, linguistics provides the organization; definitions and meanings of behavior; actions and interactions on all levels of society, business, and government, in all areas of society from individuals to families to communities, nonprofits, international groups, and more.

At this point, it is important to acknowledge some history within linguistics that led to a paradigmatic revolution—linguistics become a science due to Noam Chomsky who in the late 1950s and early 1960s made linguistics into a "science" by arguing for "formalism" through the creation of theories in "transformational grammar." The book is concluding because behind the need to stop climate change and use new technologies, there is the key "cost" or economics factor which makes an argument for the application of linguistic theory to business and economics. In short, modern linguistic theory provides the framework for a paradigm revolution in business and economics. Chomsky is the place to start since he himself started a paradigm revolution in linguistics in the late 1960s.

Most of modern economic theory is based on the seminal works of Chomsky from the late 1950s and early 1960s. Chomsky's was first published through The Hague Press in Holland because no American Press would publish his articles or books at that time. Clark experienced the same thing when he wrote about "qualitative economics" and then did some papers and books, the first a start-up book publisher (Coxmoor Press) in Oxford, UK, for their first book on the topic of Qualitative Economics (2008).

Chomsky's theories and books produced a major paradigm change in both linguistics and later (1970s) in psychology. By the early 1970s, however, some of his former students and others staged their own paradigmatic revolution. The difference in the two approaches to language studies rests primarily in what Chomsky initially defined as "competence" and "performance." For Chomsky, originally, competence or the mental processes in which the language user structured (that is "surface" and "deep" structures) the use of words, sentence, and thoughts was the key to linguistics. Frankly, Clark interested in all of this due to his interest in anthropology (related academically) linguistics.

The key in linguistics which Clark did for economics was to define the meaning of words which were the numbers, data and stats in economics. Linguistics performance or the meanings of those words, sentences, and thoughts were less important. Because of this difference, competence research turned into a new academic field of study: cognitive sciences. And the science of linguistics was (and is today) related to law. Clark started to study anthropology and law at the University of California, Berkeley (CAL), after he took courses at University of Illinois, Urbana, where a former student of Prof. Chomsky was teaching, George Lakoff. Clark went to Berkeley, following Lakoff, although he preferred MIT where Chomsky was a professor until 2018 when he retired.

However, there was more at CAL and studied linguistics, anthropology, and law which were all related to one another. His two key professors were Laura Nader (yes, the younger sister of Ralph Nader) except that she was also a PhD and liked to teach in the Anthropology Department, and not just do research. Laura created a Antrhopology-Law Center where Clark took classes too. The other professor at CAL was Herbert Blumer from whom Clark learned about "symbolic interactionism" and how individuals and groups behaved—and why. This process is what made sense for gathering of data rather than just getting numbers. Today, Focus Groups and other similar forms of data gathering are critical in every discipline and area.

George Lakoff along with his wife and Paul Kay, Charles Filmore, among others, objected strongly to the focus on statistics and data competence approach in research. Meaning was the most important part of language. They developed a new field, "cognitive science" which has grown substantially in the 1990s. Some scholars would argue that ethnomethodology itself represented the performance paradigm within linguistics since it focused upon the content or meaning of action. Without meaning, language competence and its formal structures made no sense.

Language competence was nontemporal, linear, and isolated from everyday life went the argument. Various subtheories grew around this paradigm such as discourse analysis, speech act theories, and pragmatics. By the late 1980s and early 1990s, however, Chomsky had incorporated performance and meaning into his theories. Or to put it another way, he stated that meaning had always been part of his paradigm ("government binding").

This book does not debate the paradigm changes within linguistics. Clearly the distinctions between competence and performance are significant. Critical to this discussion is the basic notion that meaning cannot be separate from structure. However, the discussion below is not aimed to review or enter into the current linguistic debate over these issues. Instead, the business cases of advanced technologies demonstrate that meaning and structure must be combined within a linguistic theoretical framework.

An argument is presented that uses linguistic formalism for describing and understanding the actions of actors in everyday business life. Linguistic formalism allows scholars and observes to see relationships in the context of micro and then macrotheories. Indeed, for that reason alone, transformational linguistic grammar theory is a very powerful theoretical paradigm. Linguistic theory provides the framework for combining meaning and structure in human interactionism. In short, linguistic theory allows the study of business economics to be scientific.

Above all, the importance of performance is paramount and acknowledged as a key element in everyday business life. It is the basis of what Blumer argues makes humans human—human beings reflect through the "generalized other" on their actions and those of others. In short, actors think; they do not just react to stimuli or events. Thus Blumer established the basis of understanding how and why actors interact with one another. His approach today would be called nonlinear and parallel processing of actions rather than a simplistic linear or causal sequence of activities and events. What Chomsky's linguistic transformational grammar does is formalizes the micro and situational analysis level to a universal and macroscientific level.

The everyday business activities and interactions of people and hence their work must be seen in their present situations which include a past, meaning, and other actors. Through interactionism, everyday business life transpires. It is constantly changing and moving. The conventional paradigm and neoclassical theorists would describe and quantify only one moment or snapshot in time of a business activity. Such a rigid perspective of business is clearly not realistic and certainly limited in any scientific sense. Statistics do not provide scientific understanding; certainly are not explanatory; and never predictive. Linguistic formalism, derived from an entirely different set of philosophical roots, and when combined with the same philosophical thought from interactionism, transforms business economics into a scientific discipline. Below, such a paradigmatic revolution is demonstrated in business economic.

The competence of human beings for language usage is interactive with their performance of language actions on a daily basis. Blumer, built upon Mead and a strong philosophical tradition that articulated this interactionism among actors. Hence transformational linguistics becomes a formalism derived from everyday interactionism to better understand business activities, the firm, and how they fit and influence larger and more universal events in society. It is the chore element in this chapter to articulate and merge these various parallel and connected perspectives into a new approach to business economics.

For example, from anthropology, there are some attempts to postulate macrotheories from micro field work studies. However, many of them follow the neoclassical economic paradigm into a focus only upon barter and exchange of goods and services at the community (micro) level. Consider his concept of "profit" or what he calls the "common sense" concept of profit, such that profit "take(s) the form of power, rank, or experience and skills..." (1962).

Profit is not simply the monetary or material gains often assumed in the economics literature. Such a common sense definition of profit derived from microeconomic studies in various community studies allows a more comprehensive and complete understanding for building macroeconomic theories, but it rests clearly in a predetermined organizational structure. "Costs" (the other side of the business ledger on a balance sheet) for most business economics is also viewed from a microeconomic level to derive macroeconomic theory.

Many of the academics consider not only the monetary and material costs for a new business venture but also the "social costs" for the entrepreneurial venture within the community. When an entrepreneurial business fails, the social costs in the community are high in terms of failure to perform; conflict among family members; failure to deliver on a promise or deal; or violation of trust of people. Often when a business venture works not work, there are investors, suppliers, and creditors who also lose. This is the conventional cause-effect economic model for business-economic activity. Entrepreneurship, therefore, encompasses the individuals starting business, but also an array of chain events or casual factors. The entrepreneurial venture must constantly "balance" the monetary and human profits and losses.

The other considerations of business and work are particularly the commercialization of an innovation, allows any observer to understand the building of macroeconomic theory from the Lifeworld tradition for everyday business life. Interactionism is the cornerstone for moving from the microeconomics or case study perspective to the larger more "universal" macroeconomic one. In that context, there is a need to understand the case about technological innovation in light of business economics. The further definition of economics as exchange and therefore as engagement allows us to understand economic phenomena, events, or situations as drawing from a knowledge base in order to construct an economic reality of everyday life. Since action among actors is predicated upon their knowledge and understanding of others, consider a description of the case in such terms.

The chart below shows how G2E needs to work and does now as noted in the book:

Areas in a home, like in businesses, government buildings, and more need to interact and communicate in groups.

Academic disciplines discover each other, only after one field cannot explain what it is doing to itself and others; or arrive at the end of long endless theoretical debates; or find phenomena impossible to describe or explain. The philosophical underpinnings of most scientific fields are rooted in inquiry, debate, and usually insights if not direct theoretical application from other fields. This is true with the physical sciences as must as it is for the social sciences and their subfields.

Language can be seen then as "A comparable approach (which) is particularly appropriate in the study of an organism whose behavior is determined by the interaction of numerous internal systems operating under conditions of great variety and complexity" (ibid.). In terms of paradigm changes and the impact on theory, Chomsky argues that "Progress in such an inquiry is unlikely unless we are willing to entertain radical idealization to construct abstract systems and to study their special properties, hoping to account for observed phenomena indirectly in terms of properties of the systems postulated and their interaction" (ibid.).

As noted above, in linguistics itself, the paradigm revolution started by Chomsky has since the 1970s been challenged and revised by other linguists. Today there are, acknowledged by Chomsky himself, three theoretical perspectives within his transformational paradigm: "standard theory," "extended standard theory," and "generative semantics" (Chomsky, 1975, p. 238). Chomsky sees himself as an "extended standard theorists" for reasons not useful for this discussion. The purpose here is to use the general "transformational grammar" theory in linguistics and explore how to apply it to build a theory in order to understand and perhaps explain new enterprises in business-economics. All of this can be and is applied to energy needs and uses for "smart grids" as well as homes and buildings.

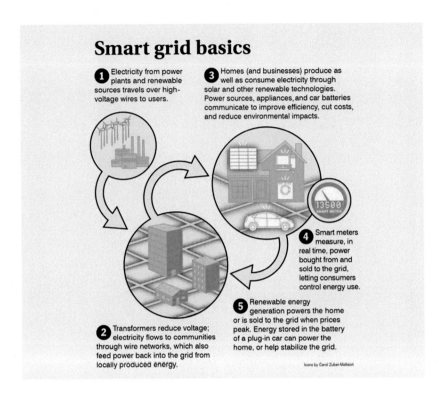

Smart grid basics

1 Electricity from power plants and renewable sources travels over high-voltage wires to users.

3 Homes (and businesses) produce as well as consume electricity through solar and other renewable technologies. Power sources, appliances, and car batteries communicate to improve efficiency, cut costs, and reduce environmental impacts.

4 Smart meters measure, in real time, power bought from and sold to the grid, letting consumers control energy use.

5 Renewable energy generation powers the home or is sold to the grid when prices peak. Energy stored in the battery of a plug-in car can power the home, or help stabilize the grid.

2 Transformers reduce voltage; electricity flows to communities through wire networks, which also feed power back into the grid from locally produced energy.

Icons by Carol Zuber-Mallison

Reproduced with permission from Environmental Defense.

The time is now to move ahead with both Q^2E and apply to solve the financial and economic problem from climate change. This is important as the need is there to watch over 100 years of fossil fuels (and now even nuclear) for power. The problem is when people and academics say "clean coal." That is a misuse of the word "clean" which with coal (and even with natural gas and nuclear) are "oxymorons." The terms are conflicting. This is critical since people and especially many companies are dependent on energy coming from coal, natural gas which are not good for the environment.

One last note on this issue to define and know the meaning of words, numbers, and data can be seen in France. The country of France has 73% of it power from nuclear power plants. The issue is that the waste from nuclear power is extremely dangerous and few people even want to talk about it. If the world and its power are to survive, the use of nuclear power needs to end soon. Green energy (solar, wind, geothermal, run of river, and hydro) is what we need to save ourselves and the planet.

USA yesterday, today, and next—the near future

14

Woodrow W. Clark II, MA³, PhD

Founder/Managing Director, Clark Strategic Partners, Beverly Hills, CA, United States

Ending this book means that Woodrow W. Clark II (Woody) needs to provide some information related to sustainable communities, the solutions to climate problems, and the costs. Clark is an internationally recognized expert, author, lecturer, and consultant on global, national, and local solutions to climate change. He has taught courses at University of International Relations in Beijing annually in July from 2013 to 2019. He was selected to be a member of the Paris Accord hosted by China in 2016 so Clark was a member of UN B20 Finance Task Force supported by China (2016)China (2016), Germany (2017), and Argentina (2018).

Clark has been very involved in both renewable energy as well as getting off of fossil fuels such as Fracking—see link below and one at the end with his work and continuing concerns over fossil fuels and nuclear energy: see below the picture of Clark being interviewed on the topic of Green Energy versus "Clean" Oil and Gas from Fracking with Clark being interviewed at UCLA in 2017 on "clean energy" that is not good as the word "clean" is not defined, but used by car companies and other groups to by-pass the pollution and climate problems from natural gas, nuclear, and coal.

Background

Clark has a core focus on economics for smart green healthy communities, where he first learned about this as a Fulbright Fellow at AAlborg University, Denmark in early 1994 and then continuing as a visiting lecturer. In 1994, Clark was recruited to be the first Manager of Strategic Planning for Energy Technology Transfer at Lawrence Livermore National Laboratory (LLNL), which was then part of the University of California and U.S. Department of Energy. In the 1990s, while at LLNL, Clark was one of the contributing scientists and researchers for United Nations Intergovernmental Panel Climate Change, which was awarded the Nobel Peace Prize in 2007. He also was the first Research Director (1996) for the UN Framework on Climate Change.

From 2000 to 2003, Clark was California Governor Gray Davis' five Energy Advisors that Clark labeled, Renewable Energy, Emerging technologies, and Finance. In 2004, Clark founded Clark Strategic Partners (CSP), a global environmental and renewable energy consulting firm. From 2004 to 2019, Clark taught courses in

Sustainable Mega City Communities. https://doi.org/10.1016/B978-0-12-818793-7.00014-7

Sustainability, Business with Economics and Technology solutions to Climate Change at Pepperdine University, University of California, Riverside and UCLA as well as universities in the EU.

Research results with 18 published books

Clark has published 16 books by the end of 2018 and 2 more by the end of 2019. He has over 80 peer-reviewed articles, which reflect his concern for global sustainable green economic healthy communities. Some of the books are listed later, but it is worth noting that his Ph.D. thesis "Violence in Schools" at University of California in 1977 was updated with a colleague into a book 4 decades later titled "Violence in Schools, Colleges and Universities" (Nova Press, 2017).

In 2018, Clark authored *Climate Preservation* (Elsevier Press) that led to three more books that are planned to be published in 2019: Qualitative Economics (second edition Springer Press) then Qualitative and Quantitative Economics (Q^{2}E): Making Economics into a Science (NOVA Press), and Sustainable Mega Cities (Elsevier). Next is Circular Economics (mid-2019). In 2014, restarted his media company that produced and distributed documentary films in the 1980s on "sexual harassment," "laughter is good medicine," and others on economics, diversity, baby boomers, and other social issues.

Aside from Clark's academic achievements and being a professor, he started four companies over his career. Woodrow (Woody) Clark's business career in environmental work started when he was 14 years old in Connecticut where he grew up as a "townie" next to Yale University. Woody and his brother (Wayne) had "worked" on a farm digging up weeds around crops, picking berries, apples, and more. They got paid $.15 an hour. Then, one day, their day wanted the boys to mow the family 1-acre yard around their home. They did that for a month. Then, the boys asked their dad to buy a sit-down lawn mower for their lawn.

Clark dad told us, "If you want a lawn mower to ride, then you boys will have to start a business and pay for it." The brothers shook hands with their dad to make a deal. The dad bought the lawn mower and we did 6—7 lawns in our neighborhood to pay for it. We were doing well. However, even better as our neighbors told their friends outside our neighborhood, so we needed a truck to drive us to and from the other lawn mowing clients. Clark was 15 years old then so could not drive a truck with the lawn mower on it.

Hence, the Clark boys' mother bought a 4-wheel drive Land Rover and a trailer that we could put the lawn mower on. She drove us to the clients who referred us to other people in their neighborhood. We called the company "Wayne—Wood Nurseries" (WWN). By the time I was 16 and driving, we were mowing over 40 lawns per week. In the winter during the holidays, we plowed snow off of drive ways and parking lots. WWN lasted a decade and paid our ways through undergraduate school and even graduate degrees until we both decided to go to University of California, Berkeley (CAL) where we both earned PhDs.

We had NO debt for college and sold WWN when we were both going to graduate schools due to making enough money and with scholarships.

Woody's second company was Clark Communications Company (CCC) founded 3 years after he earned his Ph.D. at CAL due to his making a documentary film based on the Ph.D. thesis of "Violence in Public Schools." That documentary ranked second after 60 minutes on a Sunday night. From there, Woody went on to produce and then also distribute documentary films on "social" issues in the 1980s ranging from the immigration of Jewish people to China during WWII to escape the NAZI. The Japanese controlled the eastern land of China due to winning a war with them in the early 1930s. Then, after WWII started, the Japanese saved over 10,000 from the EU and helped them live on the east coast of China. Why? The Japanese considered the Jewish people to be great inventors and successful financiers. Japan wanted Jewish people to survive.

The documentary that Clark did about the hundreds of Jewish people who left China after WWII and went to western countries like California, especially the San Francisco Bay Area to start a new life. Woody met many of these people and felt that their story needed to be told for many reasons in the early 1980s. The documentary was produced in a long (almost an hour) and released winning Clark numerous awards.

However, that was not the end of CCC as the company became more and more a distributer of social interest films and others. Clark founded a new company (his third) titled Clark Mass Media Company, which focused only on the production of media while CCC focused only on the distribution of documentaries that was a fairly new and growing business then. There were many cases of others doing that including Ted Turner with CNN and the sale of 20th Century Fox in 1985 by Rupert Murdoch after becoming an American citizen. There is much more as 20th Century Fox on December 14, 2017, and The Walt Disney Company bought the company for $52.4 billion.

All of this inspired Woody to produce more documentaries and even dramatic films that are big topics now 30 years later such as "Sexual Harassment" with Ed Asner as host, "The Healing Force" about Norman Cousins, "Older Workers" hosted by Elliot Gould, and more. These led Woody to become a member of Producers Guild of America in San Francisco, and he continues now in Los Angeles.

Finally, Woody's current company (aside from his academic teaching, research, and administration work) is Clark Strategic Partners (CSP), which was formed in early 2004 after the "recall" of California Governor Gray Davis, when Woody started a company that would focus on the solutions to climate change. CSP was the energy advisor to the LA Community College District as well as to major media studios, colleges, and building complexes. His most current critical areas today are government public policies, technologies, and economics for communities, cities, regions, and nations that all need to use and enact Circular Economics.

Clark earned three separate Master of Arts degrees from three different universities in Illinois and his Ph.D. in Cross-Disciplinary Studies at University of California, Berkeley with his Ph.D. thesis on "Violence in Schools" that he updated with a college into a book on Violence in Schools, Colleges and Universities (Nova Press, 2017)

Case from Asia: China

Now (2018) is the central focus for the Circular Economics in Asia. China and the EU signed a MOU in Beijing on July 16, 2018.

The signing of an agreement on CE by the world's two largest economies that would soon accelerate adoption of circular economy practices at a global scale. With MOU, both the EU and China could be creating the potential for a "system shift" toward a low carbon, regenerative green economy.

The key now with the UN Conference focused on the solutions to client change is for the global leaders as well as all other nations work together to see how they can manage the waste or recycling of products into viable and useful economic opportunities.

MEMORANDUM OF UNDERSTANDING ON CIRCULAR ECONOMY BETWEEN THE EUROPEAN COMMISSION AND THE NATIONAL DEVELOPMENT AND REFORM COMMISSION OF THE PEOPLE'S REPUBLIC OF CHINA

New analyses on CE by the Ellen MacArthur Foundation, published 2018, found that a transition to a Circular Economy in China's cities could make goods and services more affordable for citizens, and reduce the impacts normally associated with middle-class lifestyles, such as traffic congestion and air pollution.

The signing of an agreement on CE by the world's two largest economies that could soon accelerate adoption of circular economy practices at a global scale, creating potential for a "system shift" toward a low carbon, regenerative economy.

New analysis on CE by the Ellen MacArthur Foundation, to be published later this year, finds that a transition to a Circular Economy in China's cities could make goods and services more affordable for citizens, and reduce the impacts normally associated with middle-class lifestyles, such as traffic congestion and air pollution.

December 2018, Ellen MacArthur Foundation, Arup, Mckinsey Company & UNCTAD Report in Summary noted for China:

The circular economy opportunity for urban and industrial innovation in China

The report has 166 pages of details as to how the Chinese Economy can become circular so that communities and cities in China can become green, smart and healthy with new technologies. In the Circular Economy, buildings would be modular, durable, and flexible. The benefits of digitizing the built environment would go beyond improving energy production, efficiency, and interconnected so as to enhance sustainable productivity overall for the entire community or city.

China has had circular economy in its policy since the early 2000s. It was part of the 11th 5-year plan, and we are currently on the 13th. To begin it was primarily an industrial ecology agenda, looking at how the waste of one company can become resources for another. It was very much end of pipe, the 3 R's: reduce, reuse, and recycle. However, the latest Circular Economy Policy Portfolio, which came out in 2017, looks at ecodesign (both as a concept and as a policy) and extended producer responsibility and it is a massively important step. It shows the importance of upstream. A lot of cities are also looking at circular economy so it seems that the Chinese perception of the concept has evolved massively. Moreover, coming from a pure "how do we manage the flows" perspective, it has become an innovation agenda.

Furthermore, the Chinese economy has matured. It is not just the factory of the world, churning out cheap products. It is also an economy that is growing in investment capability, in innovation, embracing digital massively, and which has serious environmental problems that they have to deal with. All these angles converge toward a reshape of the overall system. Moreover, because they have the building blocks of a circular economy in their legislation already, they are making those gradual steps toward something that is more all-encompassing.

Graphic Design of Beijing: Here is the same as picture above with Circular Economics areas designated below in different collars yet connected.

Creating and producing as a strategy for the entire community is the need to be a circular strategy whereby protecting the environment would reshape both asset utilization and material management in the sector. The results are that people as well as the land, air, and water would all enjoy better indoor and outdoor quality of life.

Matter of priorities

Waiting to see the transposition in each Member States of the EU circular economy package is necessary to select high impact sectors with high growth potential on which Europe should bet on its near and long-term future. As for such goods with very long lifecycles (construction and infrastructure), high potential long-term investment programs can be activated. Meanwhile, in the short term, specific actions can be developed on durable consumer goods, such as electrical and electronic equipment, furniture, and vehicles. The annual turnover attributable to this type of industry is around 2600 billion euros, reflecting the importance of the recovery of raw materials that form end-of-life products (http://www.digitaleurope.org/).

Assume that the cost of raw materials, on average, contributes for 25% of the total cost of these products, then it can be expected an economic value associated with their optimal recovery, of at least 500 billion euro/year! Everyone benefits—and the roses will bloom forever. This goes beyond a Third Industrial Revolution to what is now called The Green Industrial Revolution (GIR) around the world noting many EU and Asian nations in English, 2014 (http://www.amazon.com/The-Green-Industrial-Revolution-Engineering/dp/0128023147/ref=sr_1_3?ie=UTF8&qid=14):

With an earlier version of the GIR book in Mandarin (September 2014), China sees the circular economy as a viable way to control and stop their climate problems due to greenhouse gasses and extensive pollution that impacts the entire nation—as well as the rest of the world.

The rapid transition in the United States, especially in California for example, from fossil fuel-powered cars through their combustion engines to hybrid cars to electric cars (www.tesla.com) and in 2015, to hydrogen fuel cell cars (www.hygen.com) with the creation of the various US states' hydrogen highways in less than 2 decades is significant. No longer are their combustion engines using fossil fuels.

Nor is their processing of fossil fuels for gasoline. Instead, cars are using batteries and moving rapidly into fuel cells for the energy to move cars on electricity. The circular economy has a significant role in all of this as the use of reuse of combustion engines has demised but the "smart" cars are now mostly electrically operated and functioning that means new circular economy needs for their systems.

Aside from the reduction of greenhouse gases and lowering pollution, despite a 5-year water drought, the rapid growth of all electric cars has created over 300,000 new jobs in the United States, focused mostly on California (www.latimes. nov18,2015;p1). What that means a massive change in economics and industrial growth, as the automotive business was the basic business that built many nations in the past and will do so in the future.

China is embarking on that pathway now. In their case, they are seeing the future now and embarking on "leap frogging" into it as "The Future Car" (www.economist. corporate.unit.beijing/future-car-breakfaatsymposium/November19-2015) from the film, "Back to the Future" (1989) is here today.

References

Books: current to past by Clark

Clark II, W.W. (Editor/Author), 2009. Sustainable Communities. Springer Press.

Clark II, W.W. (Editor/Author), 2010. Sustainable Development Design Handbook. Elsevier.

Clark II, W W. (Lead Author and Editor), Fall 2012. The Next Economics: Global Cases in Energy Environment and Climate Change. Springer Press.

Clark II, W.W., 2014. Global Sustainable Communities. Design Handbook. Elsevier.

Clark II, W.W. (Ed.), 2017. Sustainable Communities Design Handbook, second ed. Elsevier Press.

Clark II, W.W. (Author/Editor), 2017. Agile Energy Systems: Global Green Distributed On-site and Central Power Grid. Elsevier Press.

Clark II, W.W. (Author/Editor), 2018. Climate Preservation. Elsevier Press.

Clark II, W.W. (Author/Editor), 2019. Sustainable Mega Cities. Elsevier Press.

Clark II, W.W., Bradshaw, T., 2004. Agile Energy Systems: Global Lessons from the California Energy Crisis. Elsevier Press.

Clark II, W.W., Bonato, D., 2019. Circular Economics.

Clark II, W.W, Fast, M. (Authors/Editors), Spring 2019. Qualitative and Quantitative Economics (Q^2E): Making Economics into a Science. NOVA Press.

Clark II, W.W., Fast, M., 2019. Qualitative Economics, second ed. Springer Press.

Clark II, W.W., Grant, C., Fall 2011. Global Energy Innovations. Praeger Press.

Clark II, W.W., Grant, C., 2015. Green Industrial Revolution. Elsevier.

Clark II, W.W., Grant, C., 2016. Smart Green Cities. Routledge Press (English &Mandarin).

Clark II, W.W., Grant C, August 2014. 伍德罗•克拉克,格兰特 with Anjun Jerry JIN and Ching-Fuh LIN 库克,金安君,林清富. Green Development in China (Mandarin). Ashgate and China Electric Power Press.

Clark II, W.W., Grant C, August 2014. 伍德罗•克拉克,格兰特 with Anjun Jerry JIN and Ching-Fuh LIN 库克,金安君,林清富. Green Development in China (Mandarin). Ashgate and China Electric Power Press.

Clark II, W.W., Michael, F., 2008. Qualitative Economics: Toward a Science of Economics. Coxmoor Press updated in 2018.

Papers by current date from authors

Clark, W., June 20, 2006. Beyond de-regulation and energy crisis: creating agile sustainable communities. Long Beach Business Journal 1—4.

Clark II, W.W., December 2006. Partnerships in creating agile sustainable development communities. Journal of Cleaner Production (15), 294—302. Elsevier Press.

Clark II, W.W., 2007. Eco-efficient Energy Infrastructure Initiative Paradigm". UNESCAP, Economic Social Council, Asia, Bangkok, Thailand.

Clark II, W.W., 2007. The green paradigm shift. Cogeneration and Distributed Generation Journal 22 (2), 6—38.

Clark II, W.W., 2008. Public Policy in the People's Republic of China: Toward a Model for Planning and Regulatory Rules for the Future.

Clark II, W.W., 2008. The green hydrogen paradigm shift: energy generation for stations to vehicles. Utilities Policy Journal. Elsevier Press.

Clark, I.I., June 2009. Sustainable Communities Design Handbook. Elsevier Press, NY, pp. 9–22.

Clark II, W.W., 2009. Renewable energy systems. In: Lund, H. (Ed.), Chapter 7: Analysis: 100% Renewable Energy", Renewable Energy Systems, second ed. Elsevier Press, pp. 129–159 (updated Second Edition, 2014, pp. 185–238).

Clark II, W.W. (Editor and Author), 2009. Sustainable Communities. Springer Press (which includes chapters authored: Introduction, Cases and Conclusion).

Clark II, W.W. (Lead Co-Author) and Isherwood, W. (Co-Editor), Winter 2010. Special issue on China: environmental and energy sustainable development. Utilities Policy Journal.

Clark II, W.W. (Editor and Author), October, 2010. Sustainable Communities Design Handbook, Elsevier Press (which includes chapters authored: Introduction, Cases and Conclusion).

Clark II, W.W. (Lead Editor and Co-Author) (with Michael Intriligator), Fall 2013. Special issue "Global cases in energy, environment, and climate change: some challenges for the field of economics". Contemporary Economic Policy Journal, WEAI, California (not published).

Clark II, W.W., Fall 2014. The relationship between sustainable markets for sustainable development. In: Aras, G. (Ed.), Sustainable Markets for Sustainable Business. Gower Press (Chapter #7).

Clark II, W.W., 2014. Asia Leapfrogs Ahead into the Green Industrial Revolution. BOAO, Amsterdam (English and Mandarin).

Clark II, W.W. (Editor and Author), 2014. Global Sustainable Communities Design Handbook. Elsevier Press, which includes chapters authored Overview, Introduction. pp. 1-12 (Chapter #2).

Clark II, W.W., July 2018. Hanergy Holding Group: Research Center Institute. Report Paper Beijing, China. http://www.hanergy.com/showCar/carshow.html.

Clark II, W.W., Cooke, G. The Third Industrial Revolution (Chapter #2) in Woodrow W.

Clark II, W.W. (Editor and Author), December 2012. The Next Economics. Springer Press, which includes chapters authored: Introduction, pp. 1–20, Chapter #4: Fast, M., Clark II, W.W. Qualitative Economics: The Science needed in Economics. pp. 71–92, Chapter #7.

Clark Interview at Hanergy HQ All-Solar Powered Car, July 2, 2016. Beijing, China http://v.qq.com/x/page/o0313vci9m6.html.

Clark II, W.W., Demirag, I., 2006. "US financial regulatory change: the case of the California energy crisis", special Issue. Journal of Banking Regulation 7 (1/2), 75–93.

Clark II, W.W., Eisenberg, L., 2008. Agile sustainable communities: on-site renewable energy generation. Utilities Policies Journal 16 (4), 262–274.

Clark II, W.W., Hall, J., 2003. Journal of Cleaner Production, Special Issue on "*Environmental Innovation*" 11 (4).

Clark, W.W. II and W. Isherwood, "Energy Infrastructure for Inner Mongolia Autonomous Region: Five Nation Comparative Case Studies", Asian Development Bank, Manila, PI and PRC National Government, Beijing, PRC, 2007.

Clark II, W.W., Isherwood, W., 2007. Special Report on Energy Infrastructures in the West: Lessons Learned for Inner Mongolia, PRC. Asian Development Bank.

Clark II, W.W., Isherwood, W., 2009. Creating an energy base for inner Mongolia, China: 'the leapfrog into the climate neutral future'. Utilities Policy Journal.

Clark II, W.W., Isherwood, W., 2009. Report on "energy strategies for inner Mongolia autonomous region". Utilities Policy Journal. https://doi.org/10.1016/j.jup.2007.07.003.

Clark II, W.W., Isherwood, W., 2010. Leapfrogging energy infrastructure mistakes for inner Mongolia. Utilities Policies Journal, Special Issue.

Clark II, W.W., Isherwood, W., Winter 2010. Special issue on inner Mongolia, China: environmental and energy sustainable development. Utilities Policy Journal.

Clark II, W.W., Lund, H., 2007. Sustainable development in practice. Journal of Cleaner Production (15), 253−258.

Clark II, W.W., Lund, H., September 2006. Special issue on "sustainable development: the economics of energy and environmental production". Journal of Cleaner Production.

Clark, W., Lund, H., 2008. Integrated technologies for sustainable stationary and mobile energy infrastructures. Utilities Policy 16 (2), 130−140.

Clark II, W.W., Xing, L. Social Capitalism: China's Economic Rise. pp. 143−164 and Clark II, W.W., Conclusion: The Science of Economics. pp. 275−286.

Clark, W.W., Xing, L., 2013. The Political-Economics of the Green Industrial Revolution: Renewable Energy as the Key to National Sustainable Communities. Chapter #22. Oxford, UK.

Clark II, W.W., Yago, G., 2005. Financing the Hydrogen Highway. Public Policy, Milken Institute, Santa Monica, CA.

Clark, W.W. II and J. Rifkin et al., " "A green hydrogen economy", special issue on hydrogen, Energy Policy, Elsevier, vol.34, Fall 2006, (34) pp. 2630-2639.

Demirag, I., Iqbal, K., Clark II, W.W., 2009. The institutionalization of public-private partnerships in the UK and the nation-state of California". International Journal of Public Policy 4 (3/4), pp190−213.

Fast, M., Clark II, W., 2012. Qualitative economics—a perspective on organization and economic science. Theoretical Economics Letters 2 (2), 162−174.

Fast, M., Hertel, F., Woodrow II, W.C., 2014. Economics as a Science of the Human Mind and Interaction. In: Theoretical Economics Letters, vol. 4. Scientific Research Publishing Inc., pp. 477−487

Greenfield, J., Woodrow II, W.C., October 28, 2013. Re-make 'Made in America' community real estate construction and sustainable development. Los Angeles Review of Books.

Hector, B. M.D., W.W. Clark II, MA3, PhD., J. Schenk, Esq and Al Saavedra, MBA, "The universal ecolabel", International Journal of Applied Science and Technology, Vol. 4, No.4. Online at: 20 August 2014: 25 pages.

Khoodaruth, A., Oree, V., Elahee, M.K., Clark II, W.W., 2017. Exploring options for a 100% renewable energy system in Mauritius by 2050. Journal of Unities Policy (JUIP).

Li, X., Clark, W.W., 2009. Crises, opportunities and alternatives globalization and the next economy: a theoretical and critical review. In: Xing, L., Winther, G. (Eds.), Globalization and Transnational Capitalism. Aalborg University Press, Denmark (Chapter #4).

Lund, H., Clark, W., June 2008. Sustainable energy and transportation systems introduction and overview. Utilities Policy 16 (2), 59−62.

Lund, H., 2008. Special issue sustainable energy and transportation. Utilities Policy.

Peng, H., 2015. "Bring a piece of green for industrial revolution" (interview of Dr. WW Clark II). Energy Review 59−64.

Sun, X., Li, J., Wang, Y., Clark, W.W., October 26, 2013. China's Sovereign Wealth Fund Investments in overseas energy: the energy security perspective. Energy Policy.

Woodrow W. C. II, MA3PhD and Grant Cooke, MJ, "The Green Industrial Revolution," Elsevier Press, pp. 13-40 and Conclusion, pp. 559–570.

Bibliography

Bonato, D., 2017. Circular Economics. ReMedia Company, Milan, Italy. Remedia. http://www.consorzioremedia.it.

Bright Future for Hanergy Solar Cars. http://www.chinadaily.com.cn/business/motoring/2016-07/04/content_25960266.htm.

CHINA "CLEAN" Energy News, January 12, 2018. https://www.carbonbrief.org/china-leading-worlds-clean-energy-investment-says-report.

Chomsky, N., Reflections on Language. 1975. Pantheon Books, New York, NY. https://chomsky.info/.

Circular Economics News – Starbucks Coffee Case in Point. https://www.triplepundit.com/2018/11/starbucks-circular-economy-roadmap/.

Circular Economy and Raw Material Strategy: A Critical Challenge for Europe. The World In English: http://www.huffingtonpost.com/entry/circular-economy_2_b_7029130. In Italian: https://www.huffingtonpost.it/woodrow-w-clark-ii/economia-circolare-approccio-strategico-materi-prima-sfida-europa-mondo_b_6975304.html?1427799160.

Clark II., W.W., July 1996. A Technology commercialization model. Journal of Technology Transfer. Washington, DC.

Clark II., W.W., December 2006. Partnerships in creating agile sustainable development communities". Journal of Cleaner Production (15), 294–302. Elsevier Press.

Clark II, W.W., 2013. The Next Economics. Springer Press.

Clark, W.W., July 2 , 2016. Interviewed about the Launch of the Hanergy HQ ALL-Solar Powered Car. Beijing, China. http://v.qq.com/x/page/o0313vci9m6.html.

Clark, W.W., 2016. Interview at Hanergy Energy Company. Beijing China. http://www.hanergy.com/showCar/carshow.html.

Clark, W.W. (Author/Editor), 2017. Agile Energy Systems: Global Green Distributed On-Site and Central Power Grid. Elsevier Press.

Clark II, W.W. (Author/Editor), Fall 2017. Sustainable Communities Design Handbook, second ed. Elsevier Press.

Clark, W., 2017. Agile Energy Systems: Global Distributed On-Site and Central Grid Power, second ed. Elsevier Press https://www.elsevier.com/books/agile-energy-systems/clark/978-0-08-101760-9A.

Clark, W.W. (Author/Editor), 2018. Climate Preservation. Elsevier Press.

Clark, W.W. (Editor and Author), 2018. Climate Preservation. Elsevier Press: https://www.elsevier.com/books/climate-preservation-in-urban-communities-case-studies/clark/978-0-12-815920-0.

Clark II, W.W., 2019. Sustainable Mega Cities. Elsevier Press.

Clark, W.W., January 2019. Circular Economics: The Academic Overview. Pepperdine University Graziadio Business School.

Clark II, W.W. (Lead Editor and Co-Author with M. Intriligator), 2019. Special Issue "Global Cases in Energy, Environment, and Climate Change: Some Challenges for the Field of EU Policies and Plans on Circular Economics 2019". http://ec.europa.eu/environment/circular-economy/index_en.htm.

Clark, W.W., Bonato, D., 2015. Appendix: the European Union Circular Economy: the transition towards a better future. HuffPost (in English and Italian).

Clark, W.W., Bonato, D., March 31, 2015. Circular economy and raw materials: a critical challenge for Europe and the rest of the world". HuffPost Business.

Clark II, W.W., Bonato, D., 2015. Circular economics. HuffPost.

Clark II, W.W., Bradshaw, T., 2004. Agile Energy Systems: Global Lessons from the California Energy Crisis. Elsevier Press.

Clark II., W.W., Chung, R.K., 1997. Transfer of Publicly Funded R&D Programs in the Field of Climate Change for Environmentally Sounds Technologies (ESTs): From Developed to Developing Countries — A Summary of Six Country Studies. UN Framework Convention for Climate Change, Report. Bohen, Germany.

Clark II, W.W., Cooke, G., 2015. Green Industrial Revolution. Elsevier Press.

Clark II, W., Dan Jensen, J., 1997. Economic Models: The Role of Government in Business Development for the Reconversion of the American Economy.

Clark II, W.W., Demirag, I., December 2001. "Investment and capitalization: American firms" (ENRON). International Journal of Technology Management. Inderscience, UK.

Clark II, W.W., Demirag, I., 2005. Regulatory economic considerations f corporate governance". International Journal of Banking. Special Issue on Corporate Governance.

Clark II, W.W., Demirag, I., 2006. US financial regulatory change: the case of the California Energy Crisis. Journal of Banking Regulation 7 (1/2), 75—93.

Clark II, W.W., Fast, M., 2008. Qualitative Economics: Toward a Science of Economics. Coxmoor Press.

Clark II, W.W., Fast, M. (Authors/Editors), Fall 2018. Qualitative Economics, second ed.

Clark II, W.W., Fast, M., 2019. Economics as a Science: Qualitative and Quantitative Economics (Q^2E) Publisher to Be Selected.

Clark II, W.W., Fast, M. (Authors/Editors), Winter 2019. Qualitative Economics second ed.

Clark, W.W., Grant, C., March 2016. Smart Green Cities. Routledge Press.

Clark II., W.W., Isherwood, W., 2007. Energy Infrastructure for Inner Mongolia Autonomous Region: five nation comparative case studies. Asian Development Bank, Manila, PI and PRC National Government, Beijing, PRC.

Clark, W.W., Li, X., 2013. The Political-Economics of the Green Industrial Revolution: Renewable Energy as the key to National Sustainable Communities (Chapter #22). Oxford, UK.

Clark II., W.W., Lund, H., December 2006. Sustainable development in practice. Journal of Cleaner Production (15), 253—258.

Clark II., W.W., Paolucci, E., 1997a. Environmental Regulation and Product Development: issues for a new model of innovation". Journal of International Product Development Management.

Clark II, W.W., Paolucci, E., 1997b. An international model for technology commercialization. The Journal of Technology Transfer. Washington, D.C.

Clark, W.W., Peng, H., September 2015. Bring a piece of green for industrial revolution (Interview of Dr. WW Clark II). In: Energy Review (Mandarin), pp. 59—64.

Clark II, W.W., Rifkin, J., et al., Fall 2006. "A green hydrogen economy", special issue on hydrogen. Energy Policy 34 (34), 2630—2639. Elsevier.

Clark II, W.W., Cooke G., August 2014. 伍德罗•克拉克, 格兰特•with Anjun Jerry JIN & Ching-Fuh LIN 库克 金安君 林清富 Green Development Paradigm (Mandarin). Ashgate and China Electric Power Press.

Clark, I.I., Woodrow, W., Grant, C., March 2016. Smart Green Cities. Routledge Press. Link to University of International Relation's website about the field trip to Hanergy. That is in Mandarin. July 13, 2016. http://www.uir.cn/Root_jxkydt/567303.shtml.

Clark, I.I., Woodrow, W., Kuhn, L., 2017. Violence in Schools, Colleges and Universities. NOVA Press.

Climate Change Report, January 11, 2019. A New World China Seen as 'absolute Winner' of Clean Energy Transition https://www.euractiv.com/section/energy/news/china-seen-as-absolute-winner-of-clean-energy-transition/.

Economics, Fall, 2013. Contemporary Economic Policy Journal, WEAI, California (not published).

Electric Buses in China, 2018. https://cleantechnica.com/2018/01/06/not-just-shenzhen-jaw-dropping-china-electric-bus-roundup/.

Electric Cars, 2018. Green Car Report. https://www.greencarreports.com/news/1120235_latest-climate-study-says-its-already-too-late Green Car, 2018.

Ellen MacArthur Foundation, December 2018. Circular Economics. Programs. The Circular Economy Opportunity for Urban and Industrial Innovation in China. ARUP Report. https://www.ellenmacarthurfoundation.org/.

Ellen MacArthur Foundation, December 2018. The Circular Economy Opportunity for Urban & Industrial Innovation in China. ARUP, Prime Company Authors, pp. 1–166.

Energy News, January 2019. China: Green Energy. http://www.greenenergytimes.org/2019/01/29/january-29-green-energy-news-6/.

EV Cars in China — Tesla and More. January 2019. https://insideevs.com/china-too-many-electric-cars/.

Green B., Makower, J. 2019. Founder and Chair of Green Biz (Founded 1999). Oakland, CA. https://www.greenbiz.com/.

Green Car Reports Daily Headline News, December 27, 2018.

Hanergy Group Updates, 2016–2018. English Language Reports on Hanergy Solar Cars BEST in English.

Hanergy Launches Solar Powered Cars in China. http://www.forbes.com/sites/tychodefeijter/2016/07/04/hanergy-launches-solar-powered-cars-in-china/#803309a126fd.

Hanergy Unveils Solar-Powered Cars to Expand Use of Technology. http://www.bloomberg.com/news/articles/2016-07-02/hanergy-unveils-solar-powered-cars-to-expand-use-of-technology.

He W, Jing, S., about Shafer, Steve, 3M China Company, Minnesota. US Firms Upbeat on Long-Term Benefits. ChinaDaily (updated September 11, 2018).

Makower, J., 2019. State of Green Business. GreenBiz, Oakland, CA.

Mann, M.E., March 2019. The weather amplifier: strange waves in the jet stream foretell a future of heat waves and floods. Science American 43–49. scientificamerican.com/magazine/sa.

McDonough, W. Cradle to Cradle C2C.

McDonough, W. Cradle to Cradle (C2C) on Circular Economics 2016: Hyperlink. http://www.i-sis.org.uk/closedLoopCircularEconomy.php.

McDonough W. www.cradletocradle.com.

Ocasio-Cortez, A., Markey, E., February 2019. Green New Deal. Washington DC, Resolution to USA Congress, The Green New Deal. https://www.eenews.net/stories/1060106501.

Seifer, M.W., 1996. The Life and Times of Nicola Tesla: Biography of a Genius. Citadel Press. http://www.marcseifer.com/tesla-reviews.htm.

Tesla on History Channel — 10PM EST (Marc Seifer). Friday May 4, 2017. https://www.history.com/shows/the-tesla-files.

US House of Representatives. Green Deal Resolution: Recognizing the Duty of the Federal Government to Create a Green New Deal. October 2018.

Digital Version of Science and Technology Daily

http://digitalpaper.stdaily.com/http_www.kjrb.com/kjrb/html/2016-07/05/node_2.htm. And some videos about the event and Hanergy: http://www.hanergy.com/about/about_videos. html.

English language reports on Hanergy solar cars

Bright future for Hanergy solar cars. http://www.chinadaily.com.cn/business/motoring/2016-07/04/content_25960266.htm.

Hanergy Launches Solar Powered Cars in China. http://www.forbes.com/sites/tychodefeijter/2016/07/04/hanergy-launches-solar-powered-cars-in-china/#803309a126fd.

Hanergy Unveils Solar-Powered Cars to Expand Use of Technology. http://www.bloomberg.com/news/articles/2016-07-02/hanergy-unveils-solar-powered-cars-to-expand-use-of-technology.

Hanergy Unveils Solar Powered "Zero Charge" EVs. http://reneweconomy.com.au/2016/hanergy-unveils-solar-powered-zero-charge-evs-37897.

Ming L., June 29, 2016. http://blogs.wsj.com/chinarealtime/2015/06/22/despite-stumbleshanergy-sees-bright-future-for-solar-powered-cars/.

Here are the links of the four overseas subsidiaries of Hanergy, where you can find more information about Hanergy's technologies in English

http://www.altadevices.com/.

http://www.globalsolar.com/.

http://miasole.com/.

http://solibro-solar.com/en/home/.

B20 Task Force Reports, September 2016. http://en.b20-china.org/.

China Daily — September 5. http://www.chinadaily.com.cn/.

China Daily — 6 September China as Global Leader. https://thinkprogress.org/g20-climate-change-fc301e6f2827#.kdwkbh574.

China Daily, September 3–5, 2016. China's President XI Jinping Met with USA President Barack Obama at the UN G-20 in Hangzhou, China. http://iosnews.chinadaily.com.cn/newsdata/news/201609/04/416167/article.html.

Clark II, W.W. (Editor /Author), 2009. Sustainable Communities. Springer Press.

Clark II, W.W. (Editor/Author), 2010. Sustainable Development Design Handbook. Elsevier.

Clark II, W.W. (Lead Author and Editor), Fall 2012. The Next Economics: Global Cases in Energy, Environment, and Climate Change. Springer Press.

Clark II, W.W. (Lead Author and Chief Editor), 2014. Global Sustainable Communities Design Handbook. Elsevier Press.

Clark II, W.W., Cooke, G., August 2014. 伍德罗•克拉克,格兰特•with Anjun Jerry JIN and 库克,金安君,林清富 Green Development Paradigm (Mandarin). Ashgate and China Electric Power Press. http://www.amazon.com/The-Green-Industrial-Revolution Engineering/dp/0128023147/ref=sr_1_3?ie=UTF8&qid=14).

Clark, W.W., Hanergy Holding Group, July 2018. http://www.hanergy.com/showCar/carshow. html.

Clark II, W.W., Bradshaw, T., 2004. Agile Energy Systems: Global Lessons from the California Energy Crisis. Elsevier Press.

Clark II, W.W., Fast, M., 2008. Qualitative Economics: Toward a Science of Economics. Coxmoor Press.

Clark II, W.W., Grant, C., Fall 2011. Global Energy Innovations. Praeger Press.
Clark II, W.W., Grant, C., 2015. Green Industrial Revolution. Elsevier Press.
Daily, "Car Industry in China", December 28, 2018.
G20 Summit Report, September 2016. http://g20executivetalkseries.com.

Index

'*Note:* Page numbers followed by "f" indicate figures, "t" indicates tables and "b" indicate boxes.'

e United States

aylor Publisher Services